D0757663

ADVANCING METHODOLOGICAL THOUGHT AND PRACTICE

RESEARCH METHODOLOGY IN STRATEGY AND MANAGEMENT

Series Editors: T. Russell Crook, Jane Lê and Anne Smith

Recently published volumes:

RESEARCH METHODOLOGY IN STRATEGY AND MANAGEMENT VOLUME 12

ADVANCING METHODOLOGICAL THOUGHT AND PRACTICE

EDITED BY

T. RUSSELL CROOK

The University of Tennessee Knoxville, Department of Management, Haslam College of Business, USA

JANE LÊ

WHU – Otto Beisheim School of Management, Germany

ANNE D. SMITH

The University of Tennessee Knoxville, Department of Management, Haslam College of Business, USA

United Kingdom – North America – Japan
India – Malaysia – China

Emerald Publishing Limited
Howard House, Wagon Lane, Bingley BD16 1WA, UK

First edition 2020

Reprints and permissions service
Contact: permissions@emeraldinsight.com

British Library Cataloguing in Publication Data
A catalogue record for this book is available from the British Library

ISBN: 978-1-80043-080-8 (Print)
ISBN: 978-1-80043-079-2 (Online)
ISBN: 978-1-80043-081-5 (Epub)

ISSN: 1479-8387 (Series)

CONTENTS

LIST OF FIGURES

LIST OF TABLES

INTRODUCTION: IS IT ALL A GAME? RANKINGS, JOURNAL LISTS, AND THE CONTEMPORARY ROLE OF BOOK CHAPTERS

T. Russell Crook, Jane Lê and Anne Smith

Keywords: Journal rankings; business journals; book chapters; FT 50; introduction; business school rankings

As researchers, methodological enthusiasts, and editors, we have always had tremendous respect for the *Research Methodology in Strategy and Management* (*RMSM*) book series. Indeed, that is why we took over editing the series in 2019 (Volume 11). We started by talking to the previous editors – Donald (Don) Bergh and David (Dave) Ketchen – to find out why the series, despite its success in publishing high-quality methodological pieces and being a celebrated methodological resource, had been dormant since 2014. One key reason lay at the crux of it: *It is incredibly difficult to get academics to write book chapters.* Our career structures currently do not incentivize this format of publication (Corbett, Cornelissen, Delios, & Harley, 2014). Yet, when we sat down to further discuss this issue among ourselves – i.e., the editors on this volume of *RMSM* – we all noted that we had written and published chapters, despite the incentive structures discouraging such work.

The discussion among the current editors (Anne, Jane, and Russell) quickly turned into a broader conversation about the current higher education landscape, with research-intensive schools being increasingly focused on rankings and journal lists. As our host institutions, like many other schools, seek to climb the rankings, they have become more systematic in "scoring" faculty research productivity, such that scores are increasingly being driven by publishing articles in journals that appear on certain lists (Corbett et al., 2014). Despite this, as the current editors of the *RMSM* series, our collective feelings remained strong as to

Advancing Methodological Thought and Practice
Research Methodology in Strategy and Management, Volume 12, 1–12

ISSN: 1479-8387/doi:10.1108/S1479-838720200000012012

why we have chosen to edit the series, why we have published book chapters, and the value we see in continuing to do so today. When juxtaposed with recently documented problems in the review process, as well as in the confidence of published research more generally, we ask: Is it all just a game? The spoiler alert is that none of the editors think so; even calling the publishing process a "game" seems like it undermines what we, as researchers, do. But, with all this in mind, in this introductory chapter, we briefly describe how rankings and journal lists – or what some might think of as "high stakes games" – have, no doubt, become even more important, and, knowing this, we next offer our opinions on whether book chapters still matter. In the process, we take a walk down memory lane to describe why we believe in the *RMSM* book series, and we also highlight the contemporary role of book chapters more generally, as well as the many contributions made within this volume.

RANKINGS AND THE "STICKINESS" OF JOURNAL LISTS

Many, if not most, research-intensive business schools are obsessed with rankings. A high rank conferred by a prestigious publication, such as *Forbes* or the *Financial Times*, has long acted as a "proxy" for the quality of a school (Devinney, Dowling, & Perm-Ajchariyawong, 2008). A school's ultimate rank is based on many input factors, including student placement, faculty-to-student ratio, and research productivity, and these rankings help stakeholders make more informed decisions by helping them to more expediently assess school quality. Though many have questioned the validity of such rankings (e.g., Devinney, Dowling, & Perm-Ajchariyawong, 2008), there is no doubt that a high rank helps schools attract vital resources in the way of prospective student and employer interest, as well as alumni support and engagement (Dichev, 2008), to name a few. To be sure, rankings matter.

For many research-oriented business schools, alongside the obsession around rankings, journal lists have emerged as an increasingly important way to assess faculty research productivity expediently (Clark, Wright, & Ketchen, 2016) and have become maybe too important (Harley, 2019). Some lists, such as the *Financial Times* Top 50, describe only the "best" academic journals in business. Others, such as the Australian Business Deans Council's (ABDC's) list, capture a wider array of journals and categorize them based on perceived assessments of quality level. And, evidence suggests many schools are leaning on such lists to make important decisions about faculty summer support, as well as promotion and tenure decisions (Corbett et al., 2014). Today, there are several well-known journal lists that have attracted attention and have become influential for assessing research quality and productivity. As well, the journals on these lists are known for their "quality and rigor," excellent standards, and, as a result, for publishing important work (Corbett et al., 2014, p. 8).

The first well-documented list of business journals was popularized starting in 1990 by The University of Texas at Dallas, or what is commonly referred to as the UTD list. The goal of the list has been "to provide timely data on research productivity in leading research journals across major disciplines in business schools" (University of Texas-Dallas (UTD), 2020: online); the list comprises 24 leading journals that publish business research. This list remains among the most restrictive in evaluating business journal research and publication quality. Recognizing the need to evaluate research quality as inputs into business school overall rankings, the *Financial Times* created a less restrictive list that, until 2007, contained 40 leading business journals. In 2009, this list expanded to 45 journals, and in 2016, it expanded yet again to include 50 leading business journals. Looking at these lists (i.e., UTD and the *Financial Times*), a key takeaway is that each business journal is either listed as a "leading" journal or not.

More recently, a list has emerged to focus on the field of management in particular. This list was produced via a collaborative effort between Texas A&M and the University of Georgia (TAMUGA; tamugarankings.com), which allowed for an assessment of the top 150 research faculties in US business schools by ranking a school's faculty vis-à-vis other faculties. In creating their list, the TAMUGA effort took a more granular approach than UTD or the *Financial Times* by highlighting faculty research productivity in "eight top-tier journals in *Management*" (TAMUGA Rankings, 2020; emphasis added). The journals they highlight, such as *Academy of Management Journal*, *Organization Science*, and *Strategic Management Journal*, are "household names" to research active management faculty. Like the UTD and *Financial Times* lists, journals are either listed as "leading journals" or not listed at all.

Two other lists have also become increasingly important in evaluating business school faculty research quality and productivity over the last several years – (1) ABDC and (2) Erasmus Research Institute of Management (ERIM). These two lists take what might be considered a more inclusive approach to capturing research quality. The ABDC list was first created in 2008. By the numbers, it is the most inclusive list. Across all business areas, 2,682 journals comprise a list that "reflects currency and continues to assist business researchers to target appropriate, quality outlets for their work" (Australian Business Deans Council (ABDC), 2019). In order to be included in the list, journals had to "contain a substantive business element," meaning that over a period of three years, journals had to either feature more than 50% business-related publications and/or more than 50% of publications had to be written by business faculty. ABDC journals are categorized into four quality classifications: A* (highest category, representing the top 5–7% of journals; in 2019, N = 199); A (second highest category, representing the next 15–25% of journals; in 2019, N = 651); B (third highest category, representing the next 35–40% of journals; in 2019, N = 850); and C (fourth highest category, representing the remaining recognized quality journals; in 2019, N = 982).

The ERIM Journals List was originally an internal list, created by the ERIM "to contribute to the quality of the scientific output of ERIM and to the academic reputation of the institute" (ERIM, 2020). However, the list is now used extensively beyond the school, particularly within Europe. The ERIM list categorizes journals into sets, producing lists that cover the whole field of management research but can be filtered for various subfields. The *Primary Set* (P) contains the best journals in the field of management. Within this set of journals, there are three subsets. The first subset, referred to as *STAR Journals* (P*), consists of the "real top ones among best journals in the field... widely considered as truly distinctive" (ERIM, 2020). These journals are all listed in the ISI Journal Citation Reports (JCR). The second subset, the *Aspirant Journals* (P A), contains journals not yet listed in the ISI JCR but expected to be so in the near future. The third subset, the *Top Managerial Journals* (M*), lists managerially relevant journals, rather than "true research journals" (ERIM, 2020). The *Secondary Set* (S) contains other

> ...scientific refereed journals of a recognized academic reputation that do not reach the quality levels of the Primary Set... [but] stand for solid, mainstream work in research of management (ERIM, 2020: text in brackets added).

Operating on the basis of the "nonexclusivity" principle, ERIM explicitly leaves contributions to other journals open, stating that:

> This does not imply that no credit will be given for publications in journals that are not on the EJL... It is impossible to make an exhaustive list of all the journals in the domain of research in management. ERIM members may also publish in journals outside the core domain of research in management (ERIM, 2020).

Table 1 juxtaposes these lists to highlight that just five journals that publish management research are common across the five lists – *Academy of Management Journal, Academy of Management Review, Administrative Science Quarterly, Organization Science*, and *Strategic Management Journal.* These journal's acceptance rates range between 5 and 10%. And, despite the articles beyond those contributing solely to theoretical development (i.e., in *Academy of Management Review*) being heavily dependent on innovations in research methodologies, it is a rare occasion when articles whose primary focus is on research methodology are published in one of these journals. While rare, there are a few important exceptions, for example, Langley (1999) on process research methods and Semadeni, Withers, and Certo (2014) on dealing with endogeneity. What is also notable is that, when we look at the common journals, they have not changed much since UTD began categorizing journals in 1990. We thus conclude that these journals lists are pretty "sticky." Of course, publishing research in these journals is difficult in general. However, publishing methods-oriented contributions in these journals is even more difficult, and most management researchers would not consider the core journals featured on these lists as "research methods" outlets. So, where can you publish work wherein the primary focus is on advancements in research methodology?

The launch of *Organizational Research Methods* (*ORMs*) in 1998 provided a specific venue for publishing methods-oriented contributions over two decades ago. Although the work published in *ORM* is very important and, indeed, the journal is considered the elite journal in organizational research methods, *ORM* will not be found on most of the "top journal lists." Despite this fact, it is still very difficult to publish in this journal. Since 1998, *ORM* has, on average, published 4 issues and about 30 articles per year, with an emphasis during its first decade

Table 1. Widely Used Journal Lists to Assess Research Productivity.

	Australian Business Dean's Council (2016)	Financial Times (2016)	ERIM Journals List (2016)	UT Dallas (1990)	TAMUGA (2017)
Journal					
Academy of Management Journal	A*	Yes	P*	Yes	Yes
Academy of Management Perspectives	A	Yes (dropped 2016)	S		
Academy of Management Review	A*	Yes	P*	Yes	Yes
Administrative Science Quarterly	A*	Yes	P*	Yes	Yes
Entrepreneurship Theory and Practice	A*	Yes	P		
Journal of Business Ethics	A	Yes	P		
Journal of Business Venturing	A*	Yes	P*		
Journal of International Business Studies	A*	Yes	P*	Yes	
Journal of Management	A*	Yes (added 2016)	P*		
Journal of Management Studies	A*	Yes (added 2010)	P*		
Management Science	A*	Yes	P*	Yes	
Organization Science	A*	Yes	P*	Yes	Yes
Organization Studies	A*	Yes	P*		
Strategic Entrepreneurship Journal	A	Yes (added 2016)	P		
Strategic Management Journal	A*	Yes	P*	Yes	Yes

Notes: (1) ERIM Journals List (EJL) 2016–2021; The Primary Set (P journals) contains the best journals in the field of management; S= Secondary. A subset of the journals in the Primary Set, the so called P* journals (or STAR journals), is considered to contain the absolute best journals in the field. (2) The TAMUGA rankings track productivity in eight top-tier journals in management: Academy of Management Journal, Academy of Management Review, Administrative Science Quarterly, Journal of Applied Psychology, Organizational Behavior and Human Decision Processes, Organization Science, Personnel Psychology, and Strategic Management Journal. (3) Only 5 journals were considered outstanding on every list.

(until 2007) on more micro-oriented researchers (Aguinis, Pierce, & Culpepper, 2009). Recognizing the need for more "conversation" about more macro-oriented research methodology issues, Dave Ketchen and Don Bergh founded and launched the *RMSM* book series in 2004. In the first *RMSM* volume, the founders outlined the overarching goal of the series "to bridge the gap between what researchers know and what they need to know about [research] methodology" to nudge researchers to improve their practice and craft (Ketchen & Bergh, 2004, p. IX). Given the lack of journal space devoted to the topic of macro-oriented research methodology contributions, but their importance to publishing high-quality research, *RMSM* extended the range of possible publication outlets for papers tackling research methods issues.

Yet, book chapters are continuously being lambasted, with even junior scholars actively discouraged from pursuing them. In creating this volume, two early career researchers pulled out because they oriented toward more "incentive-based" outlets or because their PhD supervisor advised them against it. We understand that, for early career researchers, a book chapter may distract from one's dissertation or paper focus, and that energy must be used well. With limited time and resources, any focus away from hitting top journals might even be considered "career sabotage." However, given the long lead times and low success rates at top journals, "safer" outlets like book chapters can be important foundation outlets that allow scholars to build their profile while they work toward their "big hits." Of course, experienced scholars may also note that books are not always that well circulated, which means that chapters are likely to have a lower impact on a field than publishing in a journal on one of the core journal lists. And, while all of this is true, book chapters do have impact and we really do believe that they still matter.

BOOK CHAPTERS AND RESEARCH CAREERS: REFLECTIONS BY THE EDITORS

To begin to understand why it is important to publish book chapters, for those of you who like lists, as shown, the stability and stickiness of journal lists suggest that there are not that many "homes" for high-quality research methodology contributions. Along with the fact that many of the top management journals ultimately accept less than 10% of submissions, we might conclude that sometimes, you may have something to say that does not fit this restrictive set of acceptable outlets, which means that you need a different venue to find expression in. Book chapters are more likely to allow you this freedom of expression. Many authors who have published in the *RMSM* series, as well as other influential edited books (e.g., various handbook series, such as The SAGE Handbook of Process Organization Studies), have made their contributions because they had something that they believe needed to be said. And, in the spirit of responding to Ketchen and Bergh's (2004) nudge for researchers to improve their practice and craft, we believe that book chapters are and will continue to be important.

We all (i.e., Anne, Jane, and Russell) feel that book chapters helped build our careers. We think of them as individual bricks in our career wall, and at times even see them as the cement that binds these bricks together. Given that we continue to value book chapter contributions, we wanted to reflect on why this is the case in the hopes that will inspire others to also continue to seek them out as outlets! We use our own *RMSM* contributions as the jumping off point.

Anne: I've always had an interest in visuals, having noticed their relevance and impact in some of my empirical work. Indeed, I found it helpful to document some of the features of the case contexts I was immersed in through photographs. Although I had the data, there was little on visual methods in the literature at the time, and so the photographs lay dormant for some time, informing my understanding and analysis, but never really central to it. A request by the previous editors, Dave Ketchen and Don Bergh, led me to finally take the time to pursue my interest in visual methods. Writing the chapter together with my colleague

Joshua Ray, I found myself fascinated by visual methods in organizational studies. So, what originally started out as a side project (Ray & Smith, 2010: "Worth a thousand words: Photographs as a novel methodological tool in strategic management") became a passion, ultimately leading to multiple additional publications, including one in *ORM*. However, the starting point was the *RMSM* chapter – It inspired me, gave me focus, and set a deadline, ultimately giving me the space to discover and develop a new area.

Jane: I often feel like my first publication *should have* been a book chapter. It would have a gentler and more positive entry route to publication than the journal papers that I wrote myself sore on. I only wrote my first book chapter in 2009, years after I completed my PhD, due to encouragement from my wonderful postdoc supervisor – Paula Jarzabkowski. She suggested we write some chapter contributions based on our joint work in order to explore some new ideas, and foster some new research connections, that we would then use to develop more substantial journal papers. That chapter (Jarzabkowski, Matthiesen, & Van de Ven, 2009) has now been cited 229 times – A significant amount more than the three measly citations that my first ever journal paper has been able to attract! Writing book chapters was also less restrictive, both in terms of word count and creativity, so I ended up writing several methods contributions using this medium over the years (e.g. Jarzabkowski, Bednarek, & Lê, 2018; Jarzabkowski, Lê, & Spee, 2017; Lê & Jarzabkowski, 2011; Lê & Schmid, 2019). I found it a great way to open up the conversation to areas that were not receiving the focus that I thought they deserved and that remained somewhat nebulous as a result. I also really value these papers as resources in my teaching. In fact, that is why my colleague Torsten Schmid and I decided to produce our most recent paper in *RMSM*. We felt that the methods conversation at the time was starting to congeal around a few basic "templates" and we felt that this conversation did not accurately reflect what we saw in the literature in terms of the diversity that was represented in "state of the art" research. Our paper thus purposefully sought to open up a conversation that was converging. *RMSM* was the perfect outlet for that, since we could produce quite a chunky paper that would give us the room to really develop our perspective and the various "designs-in-use" that we saw within the strategy literature.

Russell: When I published my first book chapter in *RMSM*, I was finishing my PhD. Jim Combs and I had been discussing Venkatraman and Ramanujam's (1986) influential idea that *firm* performance is a multidimensional construct. At the time, there was not a "go to" source for describing or operationalizing its underlying dimensions, yet Jim had wanted to better understand how firm growth fit into the entire firm performance picture. Having had the itch to learn more about the dimensionality of firm performance, we, together with Chris Shook, set out to understand this, eventually writing the Combs, Crook, and Shook (2005) *RMSM* paper "The dimensionality of organizational performance and its implications for strategic management research." Looking back, we just had "something to say" about what is widely considered to be the most important dependent variable in strategic management research: Organizational performance. We built on Venkatraman and Ramanujam's work (1986) in an attempt to advance thinking about the implications of organizational performance being a multidimensional construct. In particular, we tried to conceptually and empirically establish three dimensions of performance involving for profit organizations – (1) accounting/financial (e.g., returns on assets), (2) stock market (e.g., Tobin's Q), and (3) growth (e.g., year-over-year sales growth). The chapter also noted that there are important conceptual and empirical differences between firm- and operational-performance outcomes, observing that operational outcomes (e.g., improved technology outcomes, such as number of new products), play important roles in shaping firm-level outcomes. Despite being "only" a book chapter, it has become a reasonably influential piece of work – attracting over 550 citations as of this writing. So, we are glad that Dave Ketchen and Don Bergh gave us the opportunity to write our chapter; it offered an opportunity for us to study a topic that had been on our minds for quite some time.

In short, we value our own book chapters and other book chapter contributions. That is why we edit this series. We believe that book chapter contributions provide opportunity and value that is different from, and complementary to, journal articles. Indeed, we know that many book chapters in *RMSM* have had tremendous impact. For instance, the paper by Ann Langley and Charazad Abdallah (2011) on "templates and turns" in qualitative research has been cited more than 350 times. The contribution on visual methods by Anne Smith and Josh Ray, and its subsequent contributions, is now often cited as "seminal" work in visual methods. There are many examples of excellent impact in book series like *RMSM*. Indeed, it is not only a high-quality outlet but also a useful entry pathway to scholars starting out in their careers. Look, for instance, at the fantastic and innovative piece that Indira Kjellstrand contributed to this volume – she was able to produce this piece of work despite being in her first job, with a heavy teaching load. Book chapters also are not restricted in acceptance rates the same way that journals are; so while the top journals often have an acceptance rate of 10% or less, book chapter acceptance tends to be closer to 50%. The key in such contributions is the impact and novelty of the idea, rather than the production of a perfectly polished piece of work. This allows access to publication for a broader range of scholars, allowing them build their profiles, while also targeting work to journals.

We find that book series like *RMSM* provide a critical outlet to new and established scholars by giving them the opportunity to more deeply explore methodology issues that (1) are not finding space in journals, (2) diverge from the norm in broaching controversial ideas, (3) delve more deeply into particular issues in ways that restrictive journal word counts and page limitations cannot allow, and (4) explore new ideas by taking hunches to the next step. These are all noble ventures. That is why, in rounding out this volume, we decided to reflect on why book chapters are an integral part of the academic career, whether they are formally recognized in institutional incentive schemes or not.

Like the organizational performance chapter just mentioned (i.e., Combs et al., 2005), which was started when one of the coauthors had not yet earned his PhD, book chapters also represent a way to help get careers started. As a colleague and former *RMSM* volume editor, Jeremy Short, likes to say, writing chapters offers an opportunity to build a PhD student's confidence and help him or her "publish up." Writing chapters, especially for refereed books, also allows the authors to receive feedback and learn to write responses to reviewers' comments. Again, these are especially important skills to develop early in one's career (Clark et al., 2016). The constructive environment of book chapter reviews, which tends to be "friendlier" than journal reviews, can also be encouraging to early career researchers. Book chapters are also meaningful ways to "put a flag in the sand" and anchor thesis work. It is also nice to be able to reference a book chapter in a dissertation or a follow-up publication. Having published some methods work can be particularly important for those working with novel methods, as such citations allows them to illustrate that their work is legitimate within the community and give reviewers additional confidence in the method.

RMSM is a particularly valuable outlet. Not only is the series known for publishing innovative and high-quality work on research methods but it has, over the years, also featured some of the most respected researchers, who have made substantial methods contributions in their careers, such as Don Bergh, Brian Boyd, Kathleen Eisenhardt, Denny Gioia, Mike Hitt, Dave Ketchen, Ann Langley, John Van Maanen, and Jeremy Short, to name but a few. Thus, it is not only a matter of featuring good work but also having work featured alongside other good work. We see this in much the same way that special issues can be attractive targets, not necessarily because of the journal quality per se but because of the cohort of scholars contributing to a particular conversation that one may meaningfully wish to be part of.

Consider also our contributions in this volume and "what they say." We feature two beautiful reflective pieces by Joe Hair (2020) and John Van Maanen (2020) that draw out meaningful lessons about "life as a methodologist." Joe Hair, an esteemed methodologist with over 185,000 citations, uses his *RMSM* piece to provide observations from his career and offer advice to other business scholars. He takes us back almost 50 years to his PhD and walks us through various stages of his life and career, reflecting on how the way we perceive research and publications has changed over that period (including the bygone era of awaiting postal responses from journals!), and how he eventually made a career out of publishing methods. Similarly, John Van Maanen "cobbled together" a confessional piece in which he works through a series of observations about his life as an ethnographer, reflecting on how his understanding of ethnography developed over time. He discusses his experiences of studying culture in the field, as well as representing culture through textwork or writing. Both of these pieces are deeply personal and deeply insightful reflections by leading scholars, who have been cited more times that most of us can ever hope to be!

In addition to these wonderful reflective pieces, we also feature cutting-edge research innovations. For instance, the chapter by Kjellstrand and Vince (2020) explores photoelicitation methods, demonstrating the potential of such methods for data generation, despite these being rarely used in management research. They skillfully introduce different types of photographs in their study, at the end of an interview, in order to complement the interview data and enhance emotional connection and recall. Reflecting on the process, they argue that photoelicitation can bring out peoples' lived experiences of the social context being investigated.

Madden, Madden, and Smith (2020) also introduce an innovative new method, showing us how to study compassion via photographic methods. In particular, delving deeply into the history and use of photographs as a research tool in the social sciences, they set forth compassion as a key focal area. Working with this review, they identify four ways in which photographic research methods can be used to extend understanding of compassion in organizations. They conclude by highlighting critical research decisions and possible concerns in implementing photographic methods.

Beorchia and Crook (2020) also bring to the fore a potential new data source for research on interorganizational relationships – the treasure trove that is Bloomberg SPLC data. They note that the Bloomberg data contain an impressive array of supply chain relationships, including the percentage of costs and revenues attributed to a firm's suppliers and customers. This database allows researchers to identify and build network relationships among buyers and suppliers and also allows researchers to uncover new and exciting theoretical terrain.

Shifting the focus to analysis, Mackey, McAllister, Maher, and Wang (2020) chart new methodological territory by offering a new and rigorous approach for understanding inflection points via meta-analyses that are context-specific. Mackey et al. also provide a guide that can be used as a road map for researchers examining curvilinear relationships via meta-analyses. Given that most meta-analytic work to date examines linear relationships, their approach offers much promise for other researchers to develop more nuanced knowledge about their relationships of interest.

The next two pieces we feature in this volume pick up on two important themes in research methods: Process research and computer-assisted qualitative data analysis. The first piece, a creative piece written around a panel conversation among leading process scholars, edited by Jane Lê, sees Raghu Garud, Paula Jarzabkowski, Ann Langley, Haridimos Tsoukas, and Andrew Van de Ven (2020) in conversation about critical issues in process research methods. The paper features reflections by these scholars, who draw out the different understandings of process and the implications of these understandings, for ways in which we conduct process research. Structured like a conversation, this novel format tries to capture some of the dynamism of the debate as the process scholars reflect on process research methods and their approach to process research methods more broadly, before answering specific questions about process research methods that allow for a crossing of perspectives.

The other piece by Paula O'Kane (2020) explores the utility of computer-assisted qualitative data analysis in a very informal and accessible way. Rather than trying to identify "best practice" or put together a "single user guide," Paula O'Kane genuinely uses her chapter to demystify the CAQDAS environment, articulate dilemmas and contradictions in using the software, and reassure researchers that their challenges are not unique to them! As she aptly states, this approach helps to

> …frame both the potential of CAQDAS tools but also the inherent difficulties of making these tools work for the qualitative researcher in ways which maintain individual and qualitative integrity and allow researchers to explore their data in a meaningful and useful way.

These meaningful contributions demonstrate that chapters have a contemporary role in the continued evolution and sharing of knowledge involving research methodology in strategy and different subfields of management and related areas of inquiry. We look forward to continuing to publish such work, despite the lack of institutional incentives to do so. We also look forward to continuing employing new formats – from reflections, over reviews and conversations, to introducing new methods. We believe that innovations in research methodology come in many shapes and forms – and they all matter. After all, publication is more than a game!

REFERENCES

Aguinis, H., Pierce, C. A., & Culpepper, S. A. (2009). Scale coarseness as a methodological artifact: Correcting correlation coefficients attenuated from using coarse scales. *Organizational Research Methods*, *12*(4), 623–652.

Australian Deans Business Council (ABDC). (2019). 2019 ABDC journal quality list. Retrieved from https://abdc.edu.au/research/abdc-journal-list/

Beorchia, A., & Crook, T. R. (2020). Bloomberg supply chain analysis: A data source for investigating the nature, size, and structure of interorganizational relationships. In T. R. Crook, J. K. Lê, & A. D. Smith, (Eds). *Research methodology in strategy and management* (Vol. 12, pp. 73–99). Bingley: Emerald Group Publishing Limited.

Clark, T., Wright, M., & Ketchen Jr., D. J. (2016). Publishing in management–exhilaration, bafflement and frustration. In *How to get published in the best management journals*. Northhampton, MA: Edward Elgar Publishing.

Combs, J. G., Crook, T. R., & Shook, C. (2005). The dimensionality of organizational performance and its implications for strategic management research. In D. Ketchen & D. Bergh (Eds.), *Research methodology in strategy and management* (pp. 259–286). San Diego, CA: Elsevier.

Corbett, A., Cornelissen, J., Delios, A., & Harley, B. (2014). Variety, novelty, and perceptions of scholarship in research on management and organizations: An appeal for ambidextrous scholarship. *Journal of Management Studies*, *51*(1), 3–18.

Devinney, T., Dowling, G. R., & Perm-Ajchariyawong, N. (2008). The Financial Times business schools ranking: What quality is this signal of quality? *European Management Review*, *5*(4), 195–208.

Dichev, I. D. (2008). Comment: The Financial Times business schools ranking: What quality is this signal of quality? *European Management Review*, *5*(4), 219–224.

ERIM. (2020). ERIM journals list (EJL) 2016–2021. Retrieved from https://www.erim.eur.nl/about-erim/erim-journals-list-ejl/

Financial Times. (2016). 50 journals used in FT research rank. Retrieved from https://www.ft.com/content/3405a512-5cbb-11e1-8f1f-00144feabdc0

Garud, R., Jarzabkowski, P., Langley, A., Tsoukas, H., Van de Ven, A., & Lê, J. K. (2020). Process research methods: A conversation among leading scholars. In T. R. Crook, J. K. Lê,& A. D. Smith, (Eds). *Research methodology in strategy and management* (Vol. 12, pp. 117–132). Bingley: Emerald Group Publishing Limited.

Hair, J. (2020). Musings on a distinguished methods career and beyond. In T. R. Crook, J. K. Lê, & A. D. Smith, (Eds). *Research methodology in strategy and management* (Vol. 12, pp. 13–24). Bingley: Emerald Group Publishing Limited.

Harley, B. (2019). Confronting the crisis of confidence in management studies: Why senior scholars need to stop setting a bad example. *The Academy of Management Learning and Education*, *18*(2), 286–297.

Jarzabkowski, P., Lê, J. K. & Spee, A. P. (2017). Taking a strong process approach to analysing qualitative process data. In A. Langley & H. Tsoukas (Eds.), *The SAGE Handbook of process organization studies* (pp. 237–251). London: SAGE Publications.

Jarzabkowski, P., Bednarek, R. & Lê, J. K. (2018). Studying paradox as process and practice: Identifying and following moments of salience and latency. In M. Farjoun, W. Smith, A. Langley, & H. Tsoukas (Eds.), *Perspectives on process organization studies: Dualities, dialectics, and paradoxes* (pp. 175–189). Oxford: Oxford University Press.

Jarzabkowski, P., Matthiesen, J. K., & Van de Ven, A. (2009). Doing which work? A practice approach to institutional pluralism. In T. Lawrence, B. Leca, & R. Suddaby (Eds.), *Doing institutional work* (pp. 284–316). Cambridge: Cambridge University Press.

Ketchen, D. K., & Bergh, D. (2004). Introduction. In D. Ketchen & D. Bergh (Eds.), *Research Methodology in Strategy and Management* (Vol. 1, pp. IX–X). San Diego, CA: Elsevier.

Kjellstrand, I., & Vince, R. (2020). A trip down memory lane: How photograph insertion methods trigger emotional memory and enhance recall during interviews. In T. R. Crook, J. K. Lê, & A. D. Smith, (Eds). *Research methodology in strategy and management* (Vol. 12, pp. 39–53). Bingley: Emerald Group Publishing Limited.

Langley, A., & Abdallah, C. (2011). Templates and turns in qualitative studies of strategy and management. In D. Bergh & D. Ketchen (Eds.), *Building methodological bridges: Research methodology in strategy andmanagement* (Vol. 6, pp. 201–235). Bingley: Emerald GroupPublishing Limited.

Langley, A. (1999). Strategies for theorizing from process data. *Academy of Management Review*, *24*(4), 691–710.

Lê, J. K., & Jarzabkowski, P. (2011). Touching data: Revelation through energetic collaboration. In J. Dutton & A. Carlsen (Eds.), *Generativity in doing qualitative research*. Copenhagen: Copenhagen Business School Press.

Lê, J. K., & Schmid, T. (2019). An integrative review of qualitative strategy research: Presenting 12 'Designs-in-Use'. In B. K. Boyd, T. R. Crook, J. K. Lê, & A. D. Smith (Eds.), *Research Methodology in Strategy and Management* (Vol. 11, pp. 115–155). Bingley: Emerald Group Publishing Limited.

Mackey, J. D., McAllister, C. H., Maher, L. P., & Wang, G. (2020). A guide for conducting curvilinear meta-analyses. In T. R. Crook, J. K. Lê, & A. D. Smith, (Eds). *Research methodology in strategy and management* (Vol. 12, pp. 101–115). Bingley: Emerald Group Publishing Limited.

Madden, T. M., Madden, L. T., & Smith, A. D. (2020). Capturing Organizational Compassion Through Photographic Methods. In T. R. Crook, J. K. Lê, & A. D. Smith, (Eds). *Research methodology in strategy and management* (Vol. 12, pp. 55–71). Bingley: Emerald Group Publishing Limited.

O'Kane, P. (2020). Demystifying CAQDAS: A series of dilemmas. In T. R. Crook, J. K. Lê, & A. D. Smith, (Eds). *Research methodology in strategy and management*, (Vol. 12, pp. 133–153). Bingley: Emerald Group Publishing Limited.

Ray, J. L., & Smith, A. D. (2010). Worth a thousand words: Photographs as a novel methodological tool in strategic management. In D. Ketchen & D. Bergh (Eds.), *Research Methodology in Strategy and Management* (Vol. 6, pp. 289–326). San Diego, CA: Elsevier.

Semadeni, M., Withers, M. C., & Certo, S. T. (2014). The perils of endogeneity and instrumental variables in strategy research: Understanding through simulations. *Strategic Management Journal*, *35*(7), 1070–1079.

TAMUGA. (2017). Texas A&M/university of Georgia rankings of management department research productivity. Retrieved from http://www.tamugarankings.com/

University of Texas-Dallas (UTD). (2020). The UTD top 100 business school research rankings. Retrieved from https://jindal.utdallas.edu/the-utd-top-100-business-school-research-rankings/

Van Maanen, J. (2020). Ethnography as craft: Observations on a fortunate career. In T. R. Crook, J. K. Lê, & A. D. Smith, (Eds). *Research methodology in strategy and management* (Vol. 12, pp. 25–38). Bingley: Emerald Group Publishing Limited.

Venkatraman, N., & Ramanujam, V. (1986). Measurement of business performance in strategy research: A comparison of approaches. *Academy of Management Review*, *11*(4), 801–814.

MUSINGS ON A DISTINGUISHED METHODS CAREER AND BEYOND*

Joe Hair

Keywords: Multivariate; quantitative; criticisms; journals; marketing; research

More than 40 years after the first edition of *Multivariate Data Analysis* (now in its 8th edition, Cengage Learning, UK, 2019), Dr. Joe F. Hair, Jr. continues to be actively engaged in doctoral education, research, and publishing.[1] With over 220,000 citations, Professor Hair is the author of many books and articles, and his methods contributions transcend his original marketing disciplinary boundary with widespread application and importance in organizational and management research. He has also authored *Essentials of Business Research Methods*, which is in its fourth edition and a more recent book on partial least squares, *A Primer on Partial Squares Structural Equation Modeling (PLS-SEM)*, as well as several marketing textbooks. In 2019, many of his coauthors and former students honored his contributions to business research in a book of commemorative essays *The Great Facilitator: Reflections on the Contributions of Joseph F. Hair, Jr. to Marketing and Business Research*. We are honored in this volume of *Research Methodology in Strategy and Management* to have Dr. Hair provide learning observations from his career and advice for business scholars today.

MY EARLY BACKGROUND

I earned my PhD at the University of Florida almost 50 years ago. My journey as a professor has been interesting, rewarding, and quite surprising. During my

*Adapted from "Meet the Methodologist" Interview with Larry Williams before Joe Hair's CARMA TALK on October 26, 2018 https://www.youtube.com/watch?v=MxiBYCdyQU0.
[1]Larry Williams and CARMA have graciously allowed us to use this video recorded interview with Meet the Methodologist, a conversation before the methodologist delivers a CARMA video talk.

Advancing Methodological Thought and Practice
Research Methodology in Strategy and Management, Volume 12, 13–24

ISSN: 1479-8387/doi:10.1108/S1479-838720200000012013

program, I learned a lot about marketing, but almost nothing about publishing and very little about research. In fact, none of my professors had ever published a journal article and the topic of publishing was seldom discussed. In my first job at the University of Mississippi, several other young colleagues had begun to recognize the need to conduct research and publish, but none of us had been trained much in that area. We were all assistant professors wanting to be successful, and we quickly developed a culture of working on articles and presenting papers at conferences. When we left town, which was often since Oxford, Mississippi was very small, the first thing we would do when we returned home was go to the College of Business to check the mail and see if there were any revise and resubmits, or perhaps, and not frequently enough in those early days, acceptances. Indeed, we were conditioned to react sort of like Pavlov's dogs, and our wives even made fun of us for being like them.

What really got me into research was my early interest in multivariate data analysis. When I say early interest, it really was not early in my doctoral program. It was actually late in my doctoral program when my fellow PhD students and I were required to learn multivariate. Learning multivariate data analysis is one of the worst memories of my doctoral program because we had no courses on this topic (which reflected the criticism of business education at that time as being too vocational and not sufficiently empirical or theoretical following the Carnegie and Ford reports).[2] As a reaction to these reports, the University of Florida, as did many other universities in a similar situation, hired several new professors to upgrade the business college curriculum to include scholarly activities and programs that involved empirical research, and particularly quantitative methods. One of the new professors hired by the University of Florida, Gainesville, was from UCLA, and he basically mandated that all the marketing doctoral students apply multivariate in our dissertation research. This requirement was quite frightening to all of us since we could hardly spell the word multivariate much less apply these techniques in our dissertation research. At that point, we had all finished our coursework, passed our comprehensive exams, and were ready to undertake our dissertation research. But we had very few quantitative tools in our toolbox.

In the days and weeks that followed, I said to myself, "Wow, I've done very good in marketing courses and now I'm going to have to use multivariate data analysis in my dissertation. By the way, I had a year to finish my dissertation and Ole Miss had said if I came there without my degree, I would earn only 50% of what they were offering me – $14,500 on nine months. I was newly married then too, so you know I had to get the dissertation done." So my worst memory is being required to learn multivariate data analysis after becoming ABD, and yet I had never had a course in that topic, and all the statistics books were full of equations and matrix algebra – neither of which I had any courses in either. The truth is, however, one of my best

[2]Gordon, R. A., & Howell, J. E. (1959). Higher education for business, Ford Foundation, New York, NY. Retrieved from http://www.questia.com/library/3137696/higher-education-for-business; Pierson, F. C. (1959). The education of American businessmen, Carnegie Foundation for the Advancement of Teaching, Stanford, CA. Retrieved from https://www.amazon.com/Education-American-Businessmen-Pierson-Others/dp/B0000CKIUD.

memories is how that one development has changed my career and really my life and initially led me to think about writing the multivariate book. While working on my dissertation, I quickly learned to love data analysis. I cannot say that I loved statistics early on, but I eventually began to love multivariate analysis and to really enjoy it. Mastering multivariate led me to doing scholarly research because I could always find someone who was interested in the theory, in the conceptual side, and my specialization was in data analysis. So that is really how I got started in my early years at Ole Miss and then continued my academic journey with a lot of colleagues, very good colleagues, at Louisiana State University, and later on at the other universities where I have been on the faculty, as well as presented many multivariate workshops in Europe, Asia, Australia, South America, and Africa.

An important consideration for young faculty, therefore, is developing a specialization or strength in an area that differentiates your skillset from others. This will enable you to identify and contribute as a coauthor on topics that make a contribution to your discipline. If your coauthor team has no differential advantages or unique skillset, it will be very hard for you to write articles that make a contribution and enable you to get published and cited by colleagues in your discipline as well as others.

MANAGING MY PUBLISHING CAREER

In terms of managing my career, I was very lucky, but I also worked hard. I invested early on in the area of multivariate and worked with my colleague from the University of Florida, Rolph Anderson, to write the first few editions of our multivariate book. The two of us recognized the opportunity in the textbook market for writing an applied multivariate book that would appeal to a broader market. All of the books on multivariate statistics at that time included equations on almost every page, and also matrix algebra. For example, the text *Multivariate Data Analysis* by Cooley and Lohnes (1971) claimed the authors were not statisticians, but rather data analysts. It also claimed "multivariate statistics held the keys that would unlock the secrets of human social behavior" – something that young scholars would certainly be interested in learning more about. But for students with limited math or statistics backgrounds, which was typical of most students in doctoral programs in those days, this text and others were still a huge challenge to read and comprehend. The applied statistical concepts were simply overwhelmed by the formulas, calculus, Greek symbols, and matrix algebra, not to mention the new and strange sounding vocabulary associated with statistics and data analysis! The other popular text of that time, *Multivariate Statistical Methods* (Morrison, 1967), claimed to be written as "an elementary source for multivariate techniques," but it was far from being elementary for virtually all business doctoral students.

The initial challenge for my doctoral cohort was to learn enough multivariate statistics to complete our doctoral dissertation. This task opened my eyes to the potential of multivariate techniques, but it also clearly identified the need for a more user-friendly approach to teaching multivariate statistics concepts and methods. After much frustration in seeking a relatively readable source to learn multivariate, and finding none, we decided to write the multivariate book. Our goal thus became to write a book that had no formulas and emphasized what the types of analysis techniques were designed to execute, how to select the correct method to achieve a particular research/statistical objective, how to properly apply the techniques, and finally how to correctly interpret the results when applying a particular technique. We almost achieved that in the first edition – there was only one formula. It explained the concept of a variate – which is a linear combination of multiple variables and considered the basic building block of all multivariate analysis – and also demonstrated in equation format how to calculate a variate.

One of the challenges of writing that first edition (1979) actually was not writing the book, but getting it published. At that time, my coauthors and I were unknown scholars on the faculty of relatively less prestigious universities located in the southern part of the United States. In addition, the field of multivariate statistics was viewed as a niche area with little potential. It took more than 6 years and a lot of persuading to find someone to publish the first edition of the book, and even then it was a small publisher new to the area of business and willing to take a risk. The first edition sold about 3,000 copies, enough to make it successful for the publisher. As interest in the field of quantitative tools expanded, the book was later acquired by a major publisher, MacMillan, and eventually by Prentice Hall, and now Cengage Learning. It was the applied approach of the book along with improved, more user-friendly software and PC technology that in combination made the book the leading text in the field today by a wide margin.

Writing the multivariate book helped me to learn data analysis, but I did not neglect the other areas. Today, my academic position is much different. I have the flexibility of working on a wide variety of topics with my doctoral students, so I can just pick up and explore new areas and really enjoy learning about emerging research opportunities. This flexibility contrasts with my earlier years, when I had to say, "no, stay focused." You know when you are focused, it helps you to stay on track, manage your time, and ultimately to publish. As an example, when I started out, I used to say I needed big blocks of time to get meaningful work done. Since then, I have learned that as few as 5 or 10 minutes is enough time for you to write a paragraph or two, and those paragraphs have then been written and they are finished.

Knowing the literature is really essential to be successful in publishing. So, in my earlier years, I also had to spend a lot of time keeping up with the literature. I am not saying I do not know the literature today, but I have the flexibility that I can follow my students' interests and learn new areas with them. A good example is in the last two or three months working with a particular doctoral student I have become interested in the field of social robotics and the impact of that in business, and particularly marketing. I am also looking at opportunities to explore this topic in management as well as other business discipline applications. In my earlier years, I probably might have had to dismiss that, but now I can learn about that topic because I am interested in this emerging area and have a lot more flexibility.

Looking back on my career, I would describe my interests other than methodology as being quite eclectic. What I would say is I have been able to be eclectic because of my methods skills. After about 10 years, I had sufficient skills in statistical methods, as long as I maintained them, that I could publish and continue to be a contributing scholar. That balance and the desire to be involved in the content or the substance side of research enabled me to pursue areas that I was interested in. But if you are a methodologist and only a methodologist, then you are limited to the situation of waiting to find a colleague who can prepare the literature and theory side and work with you. Thus, when you have skills and interests beyond methods, you can identify ideas and talk to colleagues about them. Social robotics is one example of my searching for research opportunities based on what is happening. I constantly read the business literature, not only on the scholarly side but also on business publications, *Wall Street Journal*, *Business Week*, and so forth, and I look for trends that are emerging because that is where I get my ideas. I am constantly seeking what is new and different. How are emerging innovations and issues potentially going to impact marketing, management, and scholarly research? What about the possibilities that this might open a new area of interest in research in marketing, management, or the social sciences in general?

MANAGING MY TEACHING CAREER

I think it is most unfortunate, but like most doctoral students in my day, I had no training in teaching. It was just assumed that if you get your doctorate, you know your field, and therefore you can be an effective teacher. There are so many things to educate doctoral students about, and never enough time, and so my priorities have typically been substance, theory, and methods, and too little focus on teaching my doctoral students how to teach. I think to me not emphasizing teaching has been a problem for 50 years, and it continues to be a problem. But the awareness of the need to help doctoral students become excellent teachers has improved some, with several education-focused journals now being published and scholarly conferences offering workshops on the topic – just not enough.

How did I grow? Well, you know, it was experiment and fail, and experiment and success, and I learned from both my mistakes and my successes. How did I do that? Well, over time, it was like I reflected on what worked and what did not work. I listened to others, I observed what the successful teachers were doing, I read books and articles on communicating and public speaking, and I constantly asked myself what was not working, and what is working. For young professors, my advice is to identify two or three professors you recall as being good teachers and ask yourself what made them a good teacher – what teaching methods did they use and why do you remember them as being good. Basically, I observed people, professors, and scholars, and they became mentors, but probably not true mentors as specifically defined. They were mentors in the sense that I learned from them, but they probably did not know that I was learning from them, and the exposure to them was very limited instead of an extended opportunity to learn from them. But I was able to absorb, reflect, and build upon what these outstanding teachers did and identify and adapt approaches that would work for me. As an example, I am not the type of person to tell jokes while lecturing so that was not a tactic that would work for me. But I have learned how to tell stories about my experiences that relate to an important concept and to bring that concept to life with a real world experience that students have often also been exposed to, such as a good or perhaps bad service quality experience.

That is how my teaching has evolved. I continue to be committed to investing more time in my teaching, but I do not know that I will have time because I have learned a lot about it over the years, developed a good foundation on teaching approaches, and there is always the pull of research and publishing, and also my administrative duties as Director of the PhD program. I get very good evaluations, not bragging there, but it is really because to be a good teacher you need to be sensitive to whether you are doing a good job or not doing a good job, are you communicating effectively, and to do so, you must consider the nonverbal signs of students as well as the verbal responses. You must constantly update your knowledge and monitor your presentations. In addition, ask for feedback from your students, and listen to and respond to their questions, and then help particularly the doctoral students to learn to be better teachers, since that is one of the roles they will be expected to excel in as an educator. Your school likely requires student evaluations, but too often only a few students complete them, and I find informal discussions are an excellent way to learn how effective your teaching is and what needs to be improved.

I do have some specific advice about teaching methods courses, but it likely could be useful for other types of courses. As I mentioned above, no students in my doctoral program at the University of Florida had any training in multivariate statistics, or even in advanced statistics. I tell people that I never had a course in statistics, except probability at the undergrad level, and I made a C in that course. "I promised myself I'd never take another statistics course, and I haven't." People just cannot believe it. What that has done is to motivate me to demonstrate to people the value of statistics in their career and their life in general, and how it can be useful above and beyond learning some fundamental formulas. You do need to understand what is happening in the black box and how to apply it. But even more so, you need to understand research design, how to select the correct statistical method, and then how to interpret the results, and particularly how to recognize whether initially the results are valid, or perhaps need to be examined further. That is been my focus much more so than formulas and solving equations.

When you teach methods, it is a complicated subject. There is a limit to how much most people can process in their minds and still understand the difficult concepts of statistics. When I teach it, I like to talk a little bit about the methods, what each method can do, and not do, and then give examples. A number of years ago, I wrote an opinion piece where they asked me about teaching multivariate.[3] One of the things I emphasized was in teaching include examples,

[3]Successful Strategies for Teaching Multivariate Statistics, *Proceedings*, International Conference on Teaching Statistics, Salvador, Brazil, July 2006, pp. 53–56.

examples, and more examples. We often say well this is what discriminant validity is, and we give students the definition or point it out in a Powerpoint file. But if you do not give an example, then it is often too difficult to make sense of these complex concepts. The same thing is true when you are trying to explain convergent validity and goodness of fit. You need to include examples that demonstrate and support the concepts. Those examples often come from the literature and the substantive side because that brings the concepts to life. It is the real life examples that people can relate to. Particularly individuals who lack a lot of knowledge about methods and they're finding it challenging. It's enough of a hurdle for students to just learn the basic concepts and ideas about multivariate and statistics. But it is equally important for them to be able to have an example, a practical example in a substantive area that they've been exposed to and can relate to in their everyday lives. As one example, I often discuss statistical concepts using restaurant experiences that almost all people have, such as slow service, waiters who cannot answer your questions, and food that does or maybe does not taste as you expected it to. This enables people in my classes to better understand the statistical concepts since they can apply them to their own experiences.

One of the most effective teaching strategies for multivariate, as well as research design, involves having students actually apply the techniques. To do so students must have access to their laptop or be in a computer lab. They also must have the appropriate software – in many instances, I use *SPSS* and more recently SmartPLS (www.smartpls.de), but other software is available today such as R, M-Plus, PSPP, and so forth. For the most part, selection of software depends upon what is available and you are familiar with, as there are several excellent alternatives. It also depends on the approach of the software. I prefer software packages that are point and click and drag and drop, since the focus can be on learning the multivariate technique, and not the software. But some of the open source software includes types of analyses that are not included in the major packages. As a further method of facilitating learning multivariate concepts and techniques, I include databases in all my research books that produce logical, believable results. For example, in one of my books, *Essentials of Business Research Methods* (Routledge, 4th ed., 2020), I created the Samouel's and Gino's case study that includes customer experiences that are marketing related and employee work environment concepts that are management related. Samouel's Greek Cuisine is a restaurant that competes with Gino's Italian Ristorante. Students have eaten out at restaurants many times and can relate to the situation. The two databases – one is survey data on customers' perceptions of the two restaurants and the other is a survey of employees' perceptions of the workplace environment of Samouel's Greek Cuisine. I typically show students how to apply the techniques with one of the databases and then assign them an exercise that enables them to replicate my initial examples on their own computers. Similar assignments with other databases (e.g., the HBAT database from the multivariate text) are given and students are asked to apply the appropriate techniques to solve the problem. Rather than telling the students which variables to use, the learning starts with asking them to review the database, identify the appropriate variables and statistical techniques, and then use the software to execute the solution. The assignments are graded to identify areas of concern, and later in class, they are summarized to reinforce the concepts in class, often using a group format. While the students are completing these assignments, I move through the class to clarify questions and ensure students are moving in the right direction. Students often have very limited experience or knowledge of statistical analysis and software, so I prepare handouts and Powerpoint slides that summarize examples from the textbook and articles, and the click-through sequence to obtain the results using all the statistical techniques. Students use these supporting materials while taking the class and also use them after class to obtain statistical analysis results for new research situations.

ADVICE FOR MANAGING YOUR CAREER

When I think back on my transitions in academia – from assistant, to associate and full professor, and then to a chaired position – I have continued to invest in learning new research designs and techniques and how to apply them. Even after rigorous doctoral programs, newly minted doctoral students still need to learn about the specifics of research and publishing. We simply cannot teach everything in a doctoral program that students need to learn. Newly minted doctoral students need to develop an interest in and focus on how to transition as an individual from the book learning and the concepts and the skills you learned in the doctoral program to how you actually apply them in the field. The early years are really important because that teaches you a lot about time management, and it also teaches you about what it takes to be published. You also need to identify good coauthors and partners in your research, and if a coauthor is not contributing do not be afraid to stop working with them and move on to others who invest their fair share of effort. Finally, you need to find mentors along your journey, so the early stage really is an investment in additional learning and expanding your skillset into areas where your knowledge is limited.

The second stage, as you move to associate professor, is about expanding your knowledge footprint and trying to work more broadly with other faculty, both at your university and other universities. This might be viewed as networking, but it also involves spreading the knowledge that you gained not only in your doctoral program but also in your early years as a faculty member. Then, as you evolve into a full professor or a chaired professor, you will likely be invited to present keynote addresses or workshops on your area of specialization or teach a short course at a university outside the United States. That is the time I believe you really need to be committed to serving your university community, your students, and your discipline. Success in your scholarly journey really is based upon being intellectually curious and pursuing lifelong learning. I encourage colleagues, particularly young colleagues, to pursue lifelong learning. We hear about it a lot, but it is too often overlooked. As you get to the full professor level, what you need to say is, "It's time for me to give back, and it's time for me to share with others what I have learned so that the young professors and the associate professors, and truthfully even some of the full professors can learn from your experiences." So that is how I would make sense of my transition.

To return to the theme of continuous learning, I would like to share something I have observed. Once people are out of their doctoral programs and in their careers, they often quit making a commitment to learn new methods. What I have heard is, "Well, I like multiple regression and I know it, why do I need to work with SEM, or if I know SEM then why do I need to learn hierarchical linear modeling?" And others say, "Well I've learned and applied covariance-based SEM with AMOS or LISREL for the last 30 years and it's worked well for me, why do I need to learn variance-based partial least squares SEM (PLS-SEM)?" Most of these individuals do not even know what PLS-SEM is, even though the method was developed in the 1970's and we have had excellent software for the last 20 years. In short, most scholars have been unwilling to invest even a little bit of time to learn it. This is quite surprising given it is a combination of two techniques that they have all been exposed to and almost all of them you have used – the two techniques are multiple regression and principle components analysis.[4] They say it is not useful or it is different or it is not needed because we have covariance-based SEM. But in fact, PLS-SEM objectives and its value as a research tool are very different from covariance-based SEM, and there are many areas PLS-SEM can be applied that the covariance-based SEM cannot be used. To summarize, many times researchers are not open enough to what is happening in the methodology area and are not investing enough time to become aware of and be able to say, "Wow, this might be something that would be useful to me. Either I need to learn it, or I need to find a colleague and a coauthor who can help me in terms of supporting this new methodology in my research." In fact, for the past decade, I have been writing in the area of PLS-SEM as well as other emerging methodologies. I have never stopped learning!

A commitment to learning and improving methodological skills also encompasses appreciation for other research approaches. Early in my career was when the multivariate area really began evolving, and people became excited about it. It helped us to publish, and realistically in my opinion, we forgot about qualitative. I am guilty of it just like I am sure a lot of others are. But by the time we hit the mid-1990s, or certainly the early 2000s, people began to realize that as a discipline we have been overlooking the qualitative side. In the last 15 or so years, maybe 20 in marketing and management, we began to reawaken our interest in the qualitative side and to pursue a much more balanced approach toward methods. Unfortunately, I think there are some individuals who have not caught up with that trend. One of the criticisms I have of scholarly journals these days is that too many of them continue to be too quantitative and not enough are balanced in their approach to accepting scholarly articles.

---

[4]The two techniques of ordinary least squares (OLS) multiple regression and principle components analysis (PCA) are executed simultaneously, instead of separately, and the statistical objective is predicting the maximum variance in the dependent constructs.

GLOBAL SHIFTS IN METHODS: QUANTITATIVE AND QUALITATIVE METHODS

When I was young in my career, we talked about qualitative, but then the business disciplines became very quantitative. More recently, we have moved toward mixed methods. It is kind of like I said earlier, marketing and management to me are very similar and we have been parallel, but sometimes it is kind of like competition. You know sometimes you are focusing on the idea of matching your competitor and then you think about well I need to move ahead, not just be equivalent to. In marketing and management, sometimes the management scholars have moved ahead of us and sometimes we have looked ahead of them, but to me overall, it is so great that we are now much more in the arena of mixed methods than we were for many years.

When we talk about the international market, I have had some opportunities in Europe for a number of years because they were slower than the United States in moving toward quantitative approaches and continued to focus on the qualitative side and not really learning enough about the quantitative side. They are catching up in many ways, however, and now I think that we are all in a very similar area of development and growth. If we extend that really to our more recent colleagues from Asia, they are beginning to see the value of mixed methods as well. Although they initially became interested in the quantitative side, because I think the perceptions were scholarly journals were emphasizing quantitative when accepting scholarly articles. My observations are that these days they are moving much more toward mixed methods because qualitative methods are a fundamental building block for quantitative approaches. I have seen so many people over the years that think a quantitative approach can make them successful and they do not need to learn qualitative approaches. Recently, we are moving into an era where reviewers and editors appreciate that you really need both approaches to be successful long term in a scholarly career.

How does one move to a more balanced approach between qualitative, conceptual/theoretical, and quantitative over time? First, both management and marketing now have conceptual/theoretical journals, but we need more outlets for this type of scholarly research.[5] I still strongly believe in quantitative methods and approaches, but it would be good to have more scholarly journals that focus on qualitative methods, and particularly conceptual and theoretical articles, because I think that would motivate more scholars to devote additional time to those research approaches.

Second, in the area of research, I think there still continues to be, and this is probably a result of the 30 or 40 years at least in marketing and to some extent management, that we have focused so much on methodological and quantitative approaches, that we still have too many journals that focus too much on the quantitative side, and that emphasis on the quantitative side means that the

[5]The primary conceptual/theoretical journal in management is AMR (Academy of Management Review) and in marketing the journal is AMSR (Academy of Marketing Science Review).

audiences for some journals are very small. You end up developing a small clique. As a good example, a number of years ago, I published in the *Journal of Marketing Research*, but it is so narrow today that I can hardly read it much less publish there. I might be able to find a quantitative colleague who could help me, but it is just too narrow, and truthfully, I cannot remember the last time I ran into more than one or two people who said the *JMR* is my type of journal. We need to move beyond that stage and say okay, how about we focus not only on the conceptual and theoretical but also the applied side of marketing and management ideas and concepts, so that we can speak to a much broader audience with a more meaningful message.

Third, sometimes I believe reviewers are too picky about narrow kinds of things. In marketing, we call it common methods bias, and in management, they call it mono methods. Yes, we need to be sensitive to it, but it seems to me if you are knowledgeable about this issue, and you learn how to address it on the front end of your research, that it is not the fatal flaw that it is sort of been made out to be. I am not saying it is not important, but I think for a while the issue of common methods variance became well, if you did not focus on this in your research that you invested a lot of time in, it is no good. Another example of overemphasizing a single topic or issue is when we are reflecting on research in general. I think there continues to be too much emphasis on statistical significance and not enough on the concept of power. What we need to do is say yes, statistical significance is important, but power is also important, and by the way not finding significance can also provide useful information. And in some instances, the lack of significance is just as meaningful as finding statistical significance in theoretical relationships.

CHANGING UNIVERSITIES, NEW CHALLENGES

I recently moved to the Mitchell College of Business at the University of South Alabama (USA), and the transition has been superb. I had been very involved in starting a DBA program at Kennesaw State University. It is a nontraditional doctoral program.[6] We had a lot of success and graduated 84 people in my 7

[6]Having worked in the area of nontraditional doctoral programs for almost 10 years, I have seen that more people are aware of DBAs and their research component. Many DBA programs are very comparable to a traditional doctoral program. What we have done is repackaged the content and graduate quality students who are competitive in the marketplace. Is it different? Yes, it is different, but in terms of the programs I have been involved in, the amount of face-to-face time in coursework is fairly comparable to most other doctoral programs that are the traditional doctoral programs. The students in the more recently developed DBA programs have management experience ranging from 5 to 10 years, and sometimes as much as 20 or 30 years. When you ask them about what problems do you want to investigate, they have personal experience with these problems. So their problem is not how do I come up with a topic to write papers on or how do I come up with a topic for my dissertation. It is how do I select from all the really interesting things that I have seen while in their industry career. Some of their experiences have gone right and some have gone wrong over the years, and they have to decide which one of these problems they are going to focus on in their doctoral research.

years there involved in the DBA program. We placed almost all of the graduates of the program in AACSB-accredited universities. My transition to the USA was to strengthen and enhance their DBA program, which was patterned after the one I developed at Kennesaw State. My task at USA is to work with the Mitchell College of Business to grow and expand their program. What I would say about managing this career transition, it is about learning a new system – what works here and what does not, and it is more of an administrative role. The transition to a new role at USA has been good and they are treating me very well. They have permitted me to continue to do workshops at universities around the world and to extend the global footprint of the university. They have also supported me in other ways, but it has been a struggle to keep up with publishing at the level I was at prior to this move because of the administrative component in what was initially a DBA program, but has now been converted to a PhD in business degree.

I continue to enjoy working with doctoral students, these days mostly with doctoral students in the Mitchell College of business, but to some extent former doctoral students now enjoying success in their academic career. For example, I just had one former student send me an email recently, and she said that she got her first journal publication. This is just her second year in the program. Others are presenting papers at conferences. In the area of management, we have an individual who just got recognized by the Southern Management Association for the outstanding graduate student paper, and other students presenting papers at AOM. Finally, my most recent project is writing a new book on marketing analytics (Hair, Harrison, & Ijjan, 2021) – coauthoring with a former doctoral student and a colleague I met through networking at the Academy of Marketing Science annual conference. All of these doctoral student successes are so rewarding and that is what I am committed to these days – helping my doctoral students and other young faculty learn how to be successful in an academic career. That is what this university supports, as do other universities with doctoral programs, and these efforts are working to advance doctoral education in marketing, management, and related business disciplines. It has been and continues to be a lot of fun!

REFERENCES

Babin B. B., & Sarstedt, M. (2019). *The great facilitator: Reflections on the contributions of Joseph F. Hair, Jr. to marketing and business research*. Cham: Springer Nature.

Cooley, W. W., & Lohnes, P. R. (1971). *Multivariate data analysis*. New York, NY: Wiley. doi: 10.1002/bimj.19730150413

Hair, J. F., Harrison, D., & Ijjan, H. (2021). *Essentials of marketing analytics*. New York, NY: McGraw-Hill Education.

Morrison, D. F. (1967). *Multivariate statistical methods*. London: Duxbury Press. https://trove.nla.gov.au/work/10646581

ETHNOGRAPHY AS CRAFT: OBSERVATIONS ON A FORTUNATE CAREER*

John Van Maanen

ABSTRACT

This chapter represents a personalized account of ethnography. As such, I have cobbled together a partial confessional – as they all are – out of the two penny nails of past papers, books, talks, and personal experience. I write as something of a literary strumpet whose task is to "teach myself." Meager subject it may be but, presumably, I have the requisite expertise. What I have to offer is a series of observation as to what my take on ethnography is today and how it developed over my career. It is an enlarging, booming scholarly and applied field – long escaped from its relatively insulated anthropological and sociological origins. As has become evident of late, the field has many adherents around the globe who subscribe to particular perspectives and practices that may differ in various ways from my own. However, the gist of this writing is to give an account of my own ethnographic perspective and practice which in part rests on chance and serendipity.

Keywords: Ethnography; confessional; social practice; fieldwork; textwork; happenstance

*This text is a slightly edited version of a keynote talk I presented to the Strategizing, Activities and Practice interest group at the Academy of Management Annual Meetings in Boston on August 12, 2019. I have foregone extensive referencing – particularly to my own work – to retain in this chapter something of the spirit of a podium talk.

Advancing Methodological Thought and Practice
Research Methodology in Strategy and Management, Volume 12, 25–38

ISSN: 1479-8387/doi:10.1108/S1479-838720200000012014

INTRODUCTION

Ethnography has carved out a recognizable niche in many domains – social work, public policy, market research, management studies, criminology, strategy, education, planning, folklore, media studies, design, primatology, architecture, to name but a few. Indeed, several years ago, I was asked to give a talk for the musicology department at Harvard. Such is the current attraction and spread of ethnography. As I've said elsewhere, I think such diversity is by and large invigorating and healthy, and I have no hidden agenda here to critique or restrict these perspectives. Nor do I aspire to offer up a proverbial best practice for I do not believe there is one. As I near the end of my academic tour of duty – I crossed over to emeritus status last month – I can really only speak from my own point of view worked out over the years in the *plein air* of the field and at the writing desk.

My remarks necessarily rest on areas I've spent most of my career exploring, namely ethnography and culture – with a distinctly small and uncapitalized "c" these days. Ethnography concerns the study and representation of culture as it plays out in the problems – strategic and otherwise – faced by a given group of people. There are two sides to the coin of ethnography: One side entails the study of culture usually involving fieldwork, which I take to be the personal and immersive engagement of the researcher for a sustained period of time into the lives of others. The other side is the representation, typically in writing, of what was absorbed and learned through such fieldwork. This is textwork, the depiction of culture.

It is a practice some claim to be the most humanistic of the sciences and the most scientific of the humanities, a somewhat remarkable if blurry convergence of approaches and standards. Broadly speaking, it is a storytelling institution with a good deal of scholarly legitimacy, whose works are commissioned, approved, and published by the leading scholarly institutions of the day and carries a National Science Foundation stamp of approval and, these days, usually an Institutional Review Board license to commit ethnography issued by a given university in line with their prescribed ethical standards. It claims a sort of documentary status by the fact that somebody actually ventures out beyond the ivory tower of learning and comfort to, with qualifications of course, "live with and live like" those from whom one is learning. And these matters are not up for grabs. Yet, to establish its scholarly status, ethnography is subjected to a double ordeal. Its documentary status must pass a test of epistemological veracity dictated by the research community to which it wishes to contribute while its storytelling status must undergo the trial of aesthetic appeal typical of the arts and humanities. Much admired, high quality ethnographies ring up high scores on both dimensions.

My ethnographic career proper began in 1969 with a year in the field with the Seattle Police Department. The study of street policing served as my doctoral dissertation (Van Maanen, 1972). It is often assumed that ethnographers choose field sites – like choosing mates – for life and some of them do. Even when not true – as I suspect is most often the case – site switching rarely breaks the link readers associate with the ethnographer and his or her first "tribe," usually their most admired work. Thus, Clifford Geertz is forever tied to the Balinese

(Geertz, 1972), just as Bill Whyte (Whyte, 1943) is forever bound to the street corners in the North End of Boston. At any rate, a long-lasting, cyclical involvement would mark my work with the police although I didn't plan it at the time.

As an aside, it is sobering to realize that relatively few ethnographers sustain the intensity of their research activity after they have completed a dissertation. It is – to say the least – a continuous challenge to find the energy, time, commitment, and resources necessary to engage in labor-intensive fieldwork and still be a member of reasonably good standing in a brick-and-mortar university – one of the most institutionally anchored workplaces in the world outside of North Korea. The ever-increasing trials and tribulations of the modern university – including the push for prodigious research output in the early stages of one's career, obligatory demands of teaching and the face work that goes with it, spare and shrinking research funds, unrelenting departmental and administrative duties, along with the rise of a measurement oriented audit culture that has come to typify many schools and academic departments – are hardly compatible with the unbearable (to some) slowness of ethnography.

But such future contingencies were then not on my mind when escaping from graduate school and joining up with the Sloan School at MIT. Astonishingly and most unexpectedly, I've been there ever since – with a few sabbatical years as a visiting professor at various places and some splendid research leaves here and there. Aside from numerous revisits to Seattle over a 25-year period, I've worked in the UK with coppers and detectives from Scotland Yard. I've also written some of fishermen in the northeast Atlantic and ride operators at Disneyland here, there, and everywhere. I've crossed more than a few geographic, occupational, and organizational borders. And like many other ethnographers crossed many theoretical boundaries as well from positivism to postmodernism, shamelessly borrowing analytic ideas wherever I could find them and using them wherever they seemed to fit. A style that is eclectic and pragmatic to the core, drawing on William James for succor and support (see, James, 1907/2007).

For all these years, I've defined myself as a fieldworker interested in various lines of work; an ethnographer with a circumspect and qualified realist bent – though I've played some with confessional and impressionist forms. This is to be sure an inflated, self-serving, and rather vague label. But the label is a good part of my scholarly identity, my engulfed self, my vocation, my calling. However, the meaning of this identity has developed and deepened over the years largely on the grounds of what I wrote and published.

THE CAREER UNFOLDING…

When you start publishing, it's impossible to see where the road might lead: What you will produce after 5 years? After 10 years? After 20 years? Or, gulp, after 50 years? At first, there is certainly the thrill of seeing your name in print – in this case, my name, and my laborious – yet underedited – words finally being read and, with any luck, appreciated and cited by others.

I started out publishing everywhere and anywhere I could. I say the work was "underedited" because I was publishing all over the scholarly and not-so-scholarly map. I wrote and edited several books on policing and organizational careers; I published articles in research annuals, door-stopping social science handbooks, and contributed invited essays to various collections of readings on special, theoretically driven topics. I wrote several pieces that appeared in obscure practitioner journals. I even wrote a quirky "how-to" primer (Van Maanen, 1973), a short book titled *The Process of Program Evaluation* for a public administration audience. Many of these publishing venues were – at best – lightly curated. It was a time, obviously, when there was less of a focus on what today we refer to – with a degree of anxiety and dread – as the esteemed "A-journals," thus allowing for a wider range of writing forms. The method obsessed, narrowing house styles for qualitative work in general and ethnography in particular had yet to be developed.

It did, however, give me a rather wide liberty of expression which was a terrific advantage because I could choose my topics on the basis of personal interest and find my scholarly voice without a lot a pressure from journal reviewers and tightly wound, focused, and discipline-defining editors. And, luckily, I was at MIT where the values and ethics are best expressed by six Anglo-Saxon words: "Don't tell me what to do." More importantly perhaps but largely in the background, there was – then as now – no overly specialized lexicon and privileged conceptual apparatus associated with ethnography other than an intense focus on the empirical. There are disciplinary (and subdisciplinary) preferences to be sure but they range widely and are rather fuzzy and loose. This relative freedom from formula or recipe gives an oh-so-slight literary air to ethnography leaving it open to a relatively stylized, improvised, situated form of social research.

True, I've had some experience with the formulaic, peevish, and – in Chick Perrow's (1985) memorable phrase – "asphalting journals." These are the high church journals of the day – the aforementioned A-journals – in which one is expected to make a "contribution" to the field – as if a "contribution" is an easy thing to do, like making a contribution to your local National Public Radio station, the Salvation Army, or Oxfam. And, I've published some pieces in *ASQ* following the dictates of the ferocious, take-no-prisoners house editor Linda Johanson (for some reflections on her time at *ASQ*, please see Johanson, 2007). I've published some in *Organization Science*. Stepping back a little to the more specialized or niche journals, I've also managed articles in the *Journal of Contemporary Ethnography*, a few in the *Sociology of Work and Occupations*, and a couple in *Human Organization*, an anthropology journal. In short, I've spent time in the journal submission trenches and been through the tedious, time-consuming, irksome, exasperating, and endless R & R rituals of an established academic field.

Ergo, a "contribution" was apparently made. But, in the early days at least, it's impossible to really know your work – the quality of it, your relationship to it, and what you want to – or are able – to say. All change over time. And, in terms of measured sorrow, I'm not all so sure that I really know the range and meaning of my work even now – a matter I'll elaborate on later in this essay. Yet, if there is an arc in terms of how I relate to my work at different stages in my career, it is of the following sort. In the early stages, I felt I was close to having a firm grip on the matters of my concern – be they gathering up ethnographic particulars or eliciting and tweaking the analytic notions I was using to pack up my representational materials – on socialization, on culture, on careers, on cops. This was a delusion.

There was and always is far more to learn, to understand, to take in, to discover. And while my studies have clear beginnings they now seem to have no endings. The idea that the scholarly record ends at the moment of publication is fundamentally wrong. There is always more to say. Even the demise of the fieldworker – god help us – doesn't conclude the enterprise given the inevitability of restudies – usually conducted by younger ethnographers out to check on their elders and add a new wrinkle or two to the existing ethnographic and historical base. I take this to be a rather common feature – not a bug – associated with many research careers – and not just ethnographic ones. As we read more, learn more, experience more, our confident hold on what we study and know for certain lessens, our interests broaden, our assumptions and conceits are challenged, our positions in the world change and we find that things are far more complicated and less certain than our early portraits or models would suggest. I now despair of ever getting things down pat, without doubt or hesitation.

As ethnographers, we produce a partial, fallible, and rather conjectural account of a particular parcel of reality. We can't claim that what we know is an unimpeachable truth. The cosmic secrets of the world are always slippery, foggy, just out of reach, ephemeral, playful, ever-changing, and shifty. Instead, what we think we know is a set of arguments about the meaning and significance of what we've seen, heard, and experienced in the field. All social settings are multidimensional, rich, and complex, quite able to support a range of interpretations depending on the researcher's interests. In this regard, I am reminded again of the anthropologist Marshall Sahlins' (1999) wry remark that there are only two certainties associated with an academic career: The first is that in the end we will all be wrong … and the second is that in the end we will all be dead. A successful academic career depends therefore on the second truth preceding the first. We all should keep this unsparing dictum in mind as we consider and assess our work.

True too is that as the career unfolds, this one in particular, shifts of interest occur – some paradigmatic, some topical, some theoretical or methodological. Some grow, some decline. What recognition I receive these days is tied – unsurprisingly – to my relatively recent work. My older work that was reasonably well cited at the time has fallen off the field's radar. The invitations I receive these days are mostly related to ethnographic representation and practice – asking me to review a manuscript, blurb a new book, be a discussant on a panel, deliver a paper at a small conference, or give a talk – and that includes the very words you're reading now. These are all activities – and not by any means unpleasant ones – that I've done a good deal of in recent years but differ rather substantially in this "the late show" of my career from what I was doing earlier.

This, in a sketchy, uneven, abbreviated form, is an ethnographic career unfolding. Recognizable phases and stages perhaps but I would add none well planned or even foreseen – a career formed more or less organically, exploring research opportunities as they fortuitously appeared. It is made sense of largely in retrospect by the semi-coherent narrative I've hastily whipped up here. But what of the work itself? What do I make of ethnography as a social practice recognizing that there is a considerable wingspan in the kinds of work that claim an ethnographic mantle? Much of this work I respect and admire even though it departs in various ways from my own version of "canonical" ethnographic practice.

ETHNOGRAPHY AS A SOCIAL PRACTICE

My view is that ethnography is a logic – a stance if you will – rather than a given method or any particular type of study. It names an epistemology – a way of knowing and the kind of knowledge that results. It is anything but a recipe. It involves fieldwork, headwork, textwork and results typically in a written representation of cultural understandings held by others – meanings about life, about work, about the past and imagined future, about disputes, about problems encountered, both mundane and acute. These are tied closely to a specific context and these days, usually put forth by ethnographers in text in a far more provisional and partial way than in the past.

This tentativeness is in part a result of the deep suspicion of all systems and grand narratives expressed over the past 25 or so years in the serious deconstructive, feminist, reflexive, postmodern critiques of realism – so that even the most realist ethnographer among us – unless struck down by a reader's block – has probably engaged with and drawn cautionary lessons from this literature. This is not to say that there is nothing beyond interpretation or that every form of knowledge is but an extension of power but it is to suggest that we now treat representational or truth claims rather gingerly.

As a process, ethnography is both dynamic and recursive and the encounter with the "foreign" is the very essence of ethnography. It involves observation, conversation, and participation used in sundry permutations and with different degrees of formality. We spend, for instance, a few days in the field, meander about the scene, hang out, watch what people do, talk to few people quite different from ourselves who hold ideas that in various ways that depart – often spectacularly – from our own. We learn what we can and then alter the questions we ask or the way we ask them and spend a few more days in the field and talk to more people. And on and on and on it goes – poking and prying with a purpose –but where it stops, nobody knows, at least at the outset. Those who revere standardization break out in hives when ethnographers hold forth about their craft.

Fieldwork generates both insider accounts from cultural members for what is going on and first-hand observations made on the scene. The balance between the two (or lack thereof) leads to either a quotation-driven ethnography based primarily on accounts derived largely from conversations and interviews, or a context-driven ethnography based primarily on events and participant observation. The two forms of data are not equivalent and I consider the latter – "I-witnessing" – to be something of the gold standard in ethnography. Accounts are indeed valuable in that the interest is on the actor's outlook, the way they make sense of themselves and their surroundings. They are of course rather easier to gather, require less commitment and time on the part of all those in the field, and provide more control to insiders in terms of what a fieldworker can learn. But, in a phrase, they tell us only what people say, not do. The difference between the two can be and often is substantial.

More generally, ethnography is improvisational, not procedural. It is path-dependent because we learn more about the activity, subjectivity, and intentionality of those we encounter in the field well after our work is begun and, the longer we are at it, the more we learn about what we need to learn next. Our knowledge accumulates and changes over time as we come closer to understanding the problems and perspectives – the situated practices and points of view – of the people from whom we are learning. Knowledge accumulates in large part because surprise – in some sense the Holy Grail of ethnography – is inevitable and taken seriously. When people do or say what we least expect, explanations are called for, however limited and tentative they may be at the moment. It is the imaginative reaction to surprise that fuels ethnography such that the early days in the field are often the most exciting, perhaps the most creative, periods of study since the learning curve for the ethnographer is rapidly accelerating. Yet, as I've discovered in all my work, cultural learning curves rarely if ever flatline – in part because culture itself is hardly static. There is always more to learn and surprise just around the bend. Exit is then largely arbitrary, having little to do with either theoretical or empirical saturation.

One becomes an ethnographer by going out and doing it and coming back to write it all up. This is both the rite of passage into the trade and a strongly held occupational norm. The central idea is to develop a narrative about what it is like to be someone else. It requires a removal from one's usual routines, familiar haunts, and everyday social relationships such that the world one is studying becomes "home" (if only temporarily) in which survival – or getting in, being in, and staying in the field – is itself premised on coming to see the world as do others who must live in it – certainly not with great subtlety at first but deepening with time and exposure. The idea is to get into a position where you can experience some of the same events, rituals, routines, achievements, and troubles that those you are learning from encounter – to be in a situation where you can feel the pressures they work under, decipher their intentions, recognize their problems, and come to understand the contingencies they face. Put bluntly, this often means taking the same kind of shit that they take.

Nothing much can prepare you for intensive fieldwork. If you can't figure out how to get close to the community, organization, occupation, neighborhood, or any odd group you want to study, if you can't figure out what to ask the proverbial natives, if you can't figure out how to build a certain type of rapport with often recalcitrant and always suspicious others, then it's time to think about a pleasant career in economic sociology or experimental social psychology where the so-called "data" are unlikely to be quite so cagey, to talk back, to question you the questioner.

On the textual side of the coin, ethnography also means getting some distance from the field. It seems an open secret among those who have done lengthy, away from home, fieldwork, know that "coming back" is as difficult and tough – if not tougher – than "going out." In retrospect, I understand this state of affairs for withdrawing from the field does not mean one has finished the ethnography (or, as some might say, even begun the ethnography). They've only been to the field. What awaits is endless time stretched out in front of them – to write a thesis, chapters, a book, to get articles out. In short, to make the fabled contribution.

Often overlooked is that exiting also makes one aware that a good deal of what was hidden or appeared as trivial while in the field is in fact critical to the work. In my case, certain things suddenly stood out – such as the skepticism and deep distrust I had absorbed from the street cops for the claims, justifications, excuses offered by those subjected to police attention; or the discomfort I experienced when exposed out of uniform in certain public places; or the fortifying comradery I felt when in the company of certain officers who were no longer with me. These were not apparent to me at the time – although recognizing them later when I was out of the field and figuring out how they fit into my work was a gradual and long drawn-out process.

Briefly, leaving the field makes apparent the consequences of fieldwork on the ethnographer. For me and perhaps other ethnographers as well this resembles something akin to the Stockholm Syndrome or Hostage Mentality (Bejerot, 1974) that results when one begins to slowly and often unconsciously identify with their captors. While I was definitely free to leave at any time during my fieldwork with the police, I also had identified – perhaps overidentified – with the street cops who were my teachers (and had my back) in the field. I had become as close to them as I could as a way of staying in their world, but my problem on exit was to get as far away from them as I could without leaving the planet.

What is required of course is to recognize and shake off the "insider's" point of view for something akin to an outsider's perspective that doesn't naturalize the culture. Easier said than done. In this regard, my field notes were still a comfort nonetheless if not exactly a guide. Read long after their inscription, they were revealing on many fronts. But I discovered also that I could recall from reading about the events I described in my notes the feelings those events generated and many of the contextual details that at the time I didn't think important enough to write down.

There is indeed a good deal made of the angst fieldworkers express on their return being chock full of stories but not knowing much of what to do with them. This has certainly been true for me. But, as I sort through my memories and materials, I begin to look for what I now call my "delicious ironies," my "strange quarks," my "VSTs" – "Very Special Things" – and began writing about these features of the life to be inscribed. These are marked features that represent those peculiarities, curiosities, paradoxes, surprises that one inevitably brings back with them from the field. To me, this is a list of the things that had most puzzled and perplexed me in the field and I felt almost instinctively that they cried out for analysis. George Stocking (2010) writing about his historical research on ethnography calls such things "juicy bits" and defines them as those "revealing incongruities" that cannot be left dangling.

VSTs do not make a thesis or a book or very often even an article. But, for me, they are a way to start writing. Coming back from a stay with the police and trying to sort out, for example, how the asshole label was attributed to some and not others was mind-boggling. I would ask why is this guy now sitting handcuffed in the back of our patrol car an asshole and be told by my partner that it's because he's a jerkoff, he doesn't know what the hell he's doing. When I would press my partner for why he doesn't know what he's doing, I was told it was because he is an asshole. I could take only so much of this circular reasoning in the field before I began to feel that my head would explode. It's a little like asking cultural members why they do what they do and getting back a blank stare of incomprehension or be told simply that it's "just the way we do things around here" – about as useful as asking plumbers why they use a wrench.

At any rate, by wrestling with these VSTs, I've come to the view that whatever narrative interpretation or argument I've been able to formulate is done so *a posteriori* – in the nature of an afterthought – and emerges for me primarily, although not exclusively, during the composition process taking place out of the field. This is the process by which I try to work out and propose a suggestively contextual – rather than authoritatively causal – explanation that at least aspires to be ineluctable such that a careful reader cannot casually dismiss my interpretation. This is basically craftwork.

How does all this stumbling around, rambling, back-and-forthing that goes into the craft come together in the end? How is it that a coherent research narrative can be fashioned and put forth in a persuasive way? It can't be purely a deductive procedure – using observations and interviews to test prior theory – for that would fail to capture the Columbian or exploratory spirit of ethnography. Nor can it be purely an inductive process for one cannot fruitfully amass more and more data or VSTs with no guiding ideas to inform a useful engagement with the particulars surfaced. Yet many students seem to operate this way and are eventually confronted with piles of unanalyzed interview transcripts, heaps of enumerated data, and hundreds – nay thousands – of pages of field notes. Many introductory texts grossly misrepresent the real work of analysis, confounding it with mechanical tasks of searching, counting, coding, and classifying – a process, akin to pulling a rabbit out of a hat by generating, for instance, raw or first-level codes, moving on second-level codes, leading to third-level and even fourth-level codes that purportedly aggregate and subsume earlier codes.

To my mind, a framework that helps organize the mass of field data without doing it harm comes mostly from trial and error, happenstance, and good fortune. It may take a long time to be formulated – years perhaps – but it comes largely from abductive reasoning, the pragmatic philosophy of Charles Peirce (1998). This is an analytic style that asks the researcher when confronted by a series of observations, to speculate continually on what underlying state of affairs might give rise to the observed phenomena. This seems to me to be precisely what creative field researchers do and a particularly apt way to characterize the analytic work of ethnography without diminishing it as formula – a continual cycling of observations and ideas until a satisfying, though forever in progress, correspondence is found.

I've used at various times all these strategies in coming to terms with my field materials – including, especially, my VSTs. However, I haven't made these descriptive and analytic leaps in splendid isolation, sealed off in my writing nook. I've written about textual practices at some length elsewhere (see, Van Maanen, 2011) but there remains a rather common and altogether misleading assumption held by many that writing is largely a solo act – the writer who writes alone. Suppressed by this image are all the contextual and social aspects of writing that on a short list would include such matters as the substantive and thematic choices we make, the analytic guides we follow, our reading of other writers; discussing our ideas of content and style, method and theory both in passing and seriously with, say, students and colleagues, the vital roles played by others who read our drafts (polished or unpolished) including fierce critics, friends (and foes), relatives, (daunting) thesis advisors both past and present. And through this highly social process we come to write in a language, tone, grammar, voice, genre, and figures of speech on topics that literally encode collectivity. We place our field materials in frameworks that seem to emerge out of an ongoing series of endless negotiation processes between ourselves and those others including reviewers, editors, and kindred spirits – both real and imagined – whom we grant the authority to help shape our work. We are perhaps remiss in acknowledging such sources – so numerous are they – but we surely do not write alone. If an unwritten ethnography is no ethnography at all, so too is an unread one.

This emphasis on the social context of textwork brings to light certain alterations or innovations in ethnographic tale telling. Among them – and hinted at earlier – is that the burden of ethnography – to represent culture – has become both heavier and messier and less easily located in time or space. The faith in an ethnographic holism – always something of an ethnographic fiction akin to Newton's frictionless space – has continued to retreat along with all those quaint claims of writers to have captured the "spirit" of the people, the "ethos of a university" or the "culture" of a nation or organization. While I think the trope of holism remains strong and dangerously seductive as a kind of literary suction pump or rhetorical imperative thought necessary to give a sense of closure to a study, but it seems to me there is more open-endedness to ethnography these days.

Another shift in ethnography stems from the "epistemological hypochondria" that Geertz famously suggested in 1988 had attached itself to ethnography. This seems to have spread widely and deeply throughout most ethnographic research communities, and most of us would now agree that all ethnographies owe a good deal of their persuasive power and wonder to contingent social, historical, and institutional forms. And no meta-argument, reflexivity, or navel-gazing can effectively question that contingency. Yet, those hypochondriacs that soldier on rather than taken to bed have mostly come to recognize that this sublime contingency matters little when it comes to putting ink to paper because any particular ethnography must still make its points by the same means available before the contingency was recognized and absorbed. These means are, by and large, the old ones and include the hard work of putting forth evidence, providing interpretation, inventing and elaborating analogies, invoking authorities, working through examples, marshalling one's tropes, and so on (and on). What matters is that the writing knits together a persuasive, intriguing, coherent, and presumably enlightening tale. Does the ethnography hold together? Do the empirical materials and analytic storyline fit?

The nature of ethnographic evidence, interpretation, authority, style may indeed have changed – more modestly I think than radically – but the appeal of any single work remains tied to the specific arguments made within a specific text and referenced to particular, not general, substantive, methodological, and narrative matters. The point here is that we now can assert the textuality of ethnographic facts and the factuality of ethnographic texts at the same time. The two lay in quite different domains and hence the work of ethnography goes on in much the same way as it did before textuality came into vogue because evidence (including I-witnessing) must still be offered up to support a claim in such a way that at least some readers are convinced that an author has something worth saying.

Changes in attitude and changes in reader response are of course possible and what is persuasive to one generation of ethnographers may look ludicrous to the next. And every generation, on coming of age, has some stake in showing their predecessors to be airheads. But the paradoxical characterization of the textuality and the factuality of ethnography vanishes with the realization that the practice of ethnography – as continually carried on by successive generations – does not remain the same because its facts, methods, theories, genres remain the same but because in the midst of change some audience still looks to it for the performance of a given task.

In the end, it is the ethnographer's ability to deliver verisimilitude, to persuade and convince readers that what they are reading is an authentic tale written by someone deeply familiar and knowledgeable about how things work and are done in some particular place, at some particular time, by some particular people that counts. Everything else that ethnography tries to do – to edify, to amuse, to challenge, to annoy, to startle, to theorize, to critique – rests on this.

ON CHANCE AND CONTINGENCY

My unexpected journey started in 1965 when I came to the then notably isolated, treeless, bare campus still under construction of the University of California, Irvine. A political science grad from Long Beach State, I came as a 22-year-old, relatively uninspired student ostensibly to sort out various inchoate career possibilities, avoid the draft, and perhaps hang out at the beach.

Worth noting is that I never applied to the school. In fact, I never applied to any graduate school including UCI. Wondering what I would do in the coming fall, I just more or less wandered on to the campus one day during the late summer and, by chance, met with some people who happened to be around, available, and willing to spend a little time with me. Fortunately, among those with whom I met was the young dean of the brand new social science department, Jim March. We had what I recall as a pleasant informal chat and, miraculously, I was told that very day that I could be admitted as a "designated special student" to a small social science program designed by and large by Jim. Provisions were attached of course. I had to send on my undergraduate records, I had to take the

GREs in October, and, left unsaid, I had to pass whatever classes I was to take in the coming fall. All very casual, informal, and impossible to imagine today.

Little did I know at the time that this was a daring and one-off experimental program – one without a history or disciplinary boundaries. The first year got me reading, taught me some basic probability and statistics, and left me slightly computer literate. The following year the Graduate School of Administration opened and initiated a master's degree program. Five students enrolled, myself included, and five managed to graduate two years later. Four went on the useful public and private management careers, but I figured I'd stick around and once again try a new program, this time a PhD program. Needless to say, this was a highly consequential choice for the program, as I experienced it as a cohort of one. It was nothing less than a free fall into an esoteric scholarly world that shaped the vocation and way of life I've been following ever since.

Let me pause and use this occasion of serendipity and good fortune as an example of the role of luck or chance plays in life. It plays, I would argue, a far, far greater role than most people recognize. Had I come a day or an hour earlier or later, I might not have met with Jim March or had the university been more advanced in its bureaucratic procedures, I never would have been admitted. The role of the dice certainly turned in my favor. But I realize most people don't like to hear of success – especially their own – explained away as luck.

There is a strong hindsight bias to think after the fact that an event – in this case, my career – was predictable even when it wasn't. There are no counterfactuals to tell us what would have happened had I not been so lucky at various points in my career. Of course, most of us are vividly aware of how hard we work and the difficult problems we face, but our day-to-day environment provides few reminders of how fortunate we are to have not been born in the South Bronx or some remote Appalachian hollow. Or, to put it in far more prosaic and personal terms – how fortunate we were to having met up with, say, an exceptional teacher, an influential coach, or a close friend steering us in what turned out to be the right direction.

Our personal career narratives are biased in another way as well. Events that work to our disadvantage are rather easier to recall than those that affect us in a positive way. A runner's sense of headwinds and tailwinds is helpful here. We are acutely aware and struggle against headwinds. But, when the winds are at our back, we're largely unaware of them – being pushed along by invisible and largely unheeded forces. If I were more conscious of the tailwinds at my back, such an awareness would bring to light that throughout my career I've been endowed with an advantageous, favored demographic profile and blessed with a privileged institutional position that featured light teaching loads; enormous intellectual help from patient colleagues who were – and are – supportive of my work; small classes; superb graduate students full of energy, talent, enthusiasm, and curiosity; generous leave policies; relatively easy access to grants and fellowships that allowed me to pursue my research interests wherever they led (even if straight into a writer's block).

These are indulgences, the tailwinds pushing me onward. Had I been at another institution or even at MIT some 10 or 20 years later, there might have been more pressure, sooner, to "cut to the chase" and submit to the sort of normal science dictates and the increasingly narrow and intimidating publish-or-perish disciplinary pursuits that are today so common. If so, I might well have done very different work. These intergenerational and institutional differences are real and I recognize that my career – like all careers – developed in a rather specific time and place with advantages that were and are, sadly, hardly widespread.

The biases I've highlighted here all work to overestimate our own role in whatever success we achieve, including ethnographic success. Most personal histories of the sort I am displaying here almost certainly exaggerate the casual significance of the individual. True, overlooking the part chance or luck plays in life may be (perversely) adaptive – encouraging us to work hard, carry on, and fight in the face of challenges. But, turning a blind eye to the role chance plays serves also to make personal narratives of success highly unreliable. Success stories based on an "n of one" are something of an exercise in fiction and imagination. History does not run controlled experiments. Personal histories retrospectively link rather vague and uncertain intentions to specified and realized outcomes when, in between the two, are hundreds and hundreds if not millions of unrecognized contingencies at play.

Now, with this display of requisite modesty and humility on the record – and ruefully acknowledging that authenticity and sincerity are largely matters of being taken in by one's own act – I would be remiss were I not to mention the hard work that goes into research careers and, paradoxically, the effort we put into hiding such work. Along with many others, I hold that success is largely a matter of grit or following a rule that says becoming proficient at anything takes dedication and practice – be it the arcane arts of archery, surgery, or ethnography. Yet, invariably, we seem to chalk up success in others to talent: Some have it and, alas, some don't. The cognitive bias here says that "natural talent" in the end trumps effort and hard work. One function of such a bias is to avoid ever having to say in the presence of accomplished practitioners: "There but for the grace of grit go I."

The partiality that is displayed for so-called "natural talent" encourages many of us to cover up all the laborious, time-consuming, and potentially embarrassing effort we put into getting good at what we do. For example, I still have something of deep terror of anyone seeing my half-written, grammatically flawed, poorly spelled, naïvely reasoned, cringeworthy drafts of papers of mine, thus perpetuating the myth that I am a natural at what I do – *the rightful words and "bon mots" just fly out folks, as fast as I can type.* The truth of the matter is that they come fitfully, woodenly, thus obscuring the amount of failure that goes into success. Such predilections make for confusing career advice: "Try hard enough and you can do anything but be sure to not let anyone know that you are trying so bloody hard."

And, so, we slog along, writing and revising endlessly – buoyed on by supportive commentary yet taken back by harsh criticism. I've certainly had my share of rejections but was always encouraged by my family, friends, and colleagues to keep at it, keep going, keep making corrections and improvements where I could but not grow disheartened or dejected as a result. I sometimes think that 90% of what I write is junk – and I've spent a hell of lot of time in that 90%. I once had a senior faculty member advise me long ago that I should write less and concentrate more of writing something really good. I realized then and now that I couldn't do that deliberately. The 10% that turns out well probably rests on learning something from the 90%. All writing is rewriting and sticking to it accompanied with a little luck in attracting readers makes all the difference in the world.

With these habits of heart and mind touched upon, let me as a way of closing remind you – Dear Fellow Researchers – that a good part of what I've said is based on the murky but optimistic premise that there is a structure to our world. Or, to be more specific, a structure to the work we do and the occupational, organizational, and institutional order in which it takes place. Moreover, the concepts we develop to decode this order promise to bring forth understandings that will actually help us act wisely, kindly, knowledgeably, strategically, productively in this the one life we have to live. But remember too what the disarming but playful ghost of Erving Goffman (e.g., Goffman, 1983) might say: When all is said and done and we are all properly admonished as to the infinite complexity of our social world and its ever-changing workings, the precepts, assurances, and sturdy practices of our intellectual fields appear rather fragile and weak and thus we must always return to the flickering, messy, cross-purposed, unknown and unknowable surroundings and circumstances that our empirical and analytic work seeks to tame. Words – and the concepts they carry – will hardly hold back the wind. But we continue – as I will – to try.

REFERENCES

Bejerot, N. (1974). The Six Day War in Stockholm. *New Scientist*, *61*(886), 486–487.

Geertz, C. (1972). Deep play: Notes on the Balinese cockfight. *Dædalus*, *101*(1), 1–37.

Geertz, C. (1988). *Work and lives: The anthropologist as an author*. Stanford, CA: Stanford University Press.

Goffman, E. (1983). The interaction order. *American Sociological Review*, *48*(1), 1–17.

James, W. (1907/2003). *Pragmatism*. New York, NY: Barnes & Noble.

Johanson, L. M. (2007). Sitting in your reader's chair: Attending to your academic sensemakers. *Journal of Management Inquiry*, *16*(3), 290–294.

Peirce, C. S. (1998). The essential Peirce. In The Peirce edition Project (Ed.), *Selected philosophical writings* (Vol. 2, 1893–1913). Bloomington, IN: Indiana University Press.

Perrow, C. B. (1985). Journaling careers. In L. L. Cummings & P. Frost (Eds.), *Publishing in organizational science*. San Francisco, CA: Jossey-Bass.

Sahlins, M. D. (1999). *Waiting for Foucault?* Cambridge: Prickly Pear Press.

Stocking, G. W. (2010). *Glimpses into my own Black Box; an exercise in self-deconstruction*. Madison, WI: University of Wisconsin Press.

Van Maanen, J. (1972). *Pledging the police: A study of Selected aspects of Recruit socialization in a large, Urban Police Department*. PhD Dissertation, University of California, Irvine.

Van Maanen, J. (1973). *The process of program evaluation*. Washington, DC: National Training and Development Service Press.

Van Maanen, J. (2011). *Tales of the field: On writing ethnography* (2nd ed.). Chicago, IL: University of Chicago Press.

Whyte, W. F. (1943). *Street corner society: The social structure of an Italian slum*. Chicago, IL: University of Chicago Press.

A TRIP DOWN MEMORY LANE: HOW PHOTOGRAPH INSERTION METHODS TRIGGER EMOTIONAL MEMORY AND ENHANCE RECALL DURING INTERVIEWS

Indira Kjellstrand and Russ Vince

ABSTRACT

The purpose of this chapter is to explore the potential of photo-elicitation as a data generating method. Photo-elicitation is rarely used for data generation, despite the considerable promise of this method. Our empirical investigation focused on people's emotions and experiences of dual systems in Kazakhstan, a country currently undergoing change from the old Soviet system to a new market economy. In addition to semistructured interviews, we use photographs in order to enhance emotional connection and recall. We use the imagery as a device to generate data, and more specifically, data on individual and social perspectives that are integral to particular experiences. We argue that photo-elicitation can bring out peoples' lived experiences of the social context being investigated. We explain why and how to use the method in practice.

Keywords: Photo-elicitation; visual research methods; emotion in organizations; semistructured interviews; projection; Kazakhstan

INTRODUCTION

In organization studies, there is an emerging interest in visual methods, and an associated body of work that illustrates the power of this approach (Boxenbaum, Jones, Meyer, & Svejenova, 2018; Davison, McLean, & Warren, 2012, 2015;

Advancing Methodological Thought and Practice
Research Methodology in Strategy and Management, Volume 12, 39–53

ISSN: 1479-8387/doi:10.1108/S1479-838720200000012015

Meyer, Höllerer, Jancsary, and Van Leeuwen (2013); Ray & Smith, 2011, 2012; Warren, 2009, 2005). While there are several different methods, including the use of drawings and video ethnography (Jarrett & Liu, 2018; Vince & Warren, 2012), our focus is on photo-elicitation and its potential as a data-generation device. Photographs are used as part of data collection, but they are mainly valued for what images depict or what the viewer interprets while looking at them (Ray & Smith, 2012; Vince & Warren, 2012; Warren, 2005). There is currently very little information for researchers on how to use photos within the research process, and on the different aspects of data they are able to generate, for example, in comparison to interviews.

In this chapter, we discuss a study that initially employed semistructured interviews and then, at the end of each interview, used photo-elicitation to engage participants in a dialogue (see also Warren, 2002). This provided us with an opportunity to consider the data available both before and after the use of this method, and to offer comparisons between the lived experiences that were communicated through verbal and visual methods. Meyer, Höllerer, Jancsary, and Van Leeuwen (2013) argue that approaches that encourage dialogue are underused and that existing studies tend to be limited to methodological papers. We address this shortcoming by using an empirical study to demonstrate the potential of photo-elicitation as a data-generating device. We found that photo-elicitation significantly enabled respondents' expression of their emotional experience and transformed the interviews.

We outline our approach to photo-elicitation, illustrating the value of the method using data from a study of peoples' lived experience and orientation toward organizational change in Kazakhstan. We discuss the implications of such an approach, focusing on its possible contributions to research in management and organization studies. We are making a methodological contribution by showing how to use photos to enrich qualitative research designs based on interviews. The rest of this chapter is organized as follows. We provide a short review of the literature on photo-elicitation methods and we explain our development of the method. We describe our example of using this method in detail and explain why we decided to structure our data collection in this way. Our findings illustrate the value of the approach by comparing the verbal and visual data. We finish this chapter by reflecting on the implications of the approach and what it can add to organizational research methods.

LITERATURE REVIEW

Using both visual and verbal methods as part of qualitative research interviews can improve the richness of the data (Boxenbaum et al., 2018). Photographs have been used to study organizations, through the semiotic study of photographs or the elicitation of responses to images (Davison et al., 2012, 2015). The former relates to what a photograph depicts and how it reflects social reality (Meyer et al., 2013), whereas with the latter, the focus is on eliciting respondents' reflection on a photograph, providing access to what people make of the image in front of them (Ray & Smith, 2012). In this chapter, our focus is on photo-elicitation – how photographs can be utilized as a tool to elicit different but complementary data alongside semistructured interview questions.

Photo-elicitation is a method of data collection that involves introducing a photographic image during the course of an interview (Harper, 2002). The approach is used to bring out *associative* aspects of peoples' lived experience, in addition to their descriptions of experience that arise from interviews (Sievers, 2008). The idea of associative experience recognizes

> …that we are all part of a matrix of relations in a social group, where certain ways of perceiving reality are impressed on the individuals without proper conscious awareness of that influence (Stamenova & Hinshelwood, 2018, p. 2).

Our sense of self in organizations is formed both knowingly and unknowingly, and our associations with an image can stimulate the articulation of emotions and relations that are characteristic of a social context.

Different Sources of Photographs

The researcher or research participant, or both, may produce photographs used in photo-elicitation; they may be archival images; or a combination of archival and "live" images (Ray & Smith, 2012). *Participant-produced photographs* are used to provide a strong voice to the research participant. The researcher asks respondents to take photos specifically for the interview, at which they are then discussed (Vince & Warren, 2012). This method is known as "autodriving" (Hurworth et al., 2005; Warren, 2005) because it is the respondent who chooses and produces the image, as well as shaping the explanation and discussion of images, and the experiences and reflections they evoke. In this way, the method emphasizes the vision and voice of the research participant. For example, the method has been used to examine "how it feels to work here" (Warren, 2002), and to uncover persons' inner experience and perception of an organization (Sievers, 2008). This particularly relates to groups of people whose voices are silenced: inmates in a prison (Sievers, 2013), homeless people (Padgett, Smith, Derejko, Henwood, & Tiderington, 2013), people undertaking "dirty work" (Slutskaya, Simpson, Hughes, Simpson, & Uygur, 2016), or those who suffer illness (Radley & Taylor, 2003).

Researcher-produced photographs involve taking images that are subsequently used as an element of an interview. The images may represent aspects of the organization that is being studied or be associated directly with research participants. For example, Petersen and Østergaard (2003) studied knowledge sharing via photographs that depicted objects such as office facilities and workspaces. They asked employees to capture the knowledge sharing processes that were happening in their organization. Heisley and Levy (1991) investigated the consumer behavior of respondents who had taken photographs during family dinners prior to the interviews. The photographs later helped respondents recall and remember product-related associations, to see signs of role behavior, power and conflicts associated with roles, and to visualize the power relations between the participants and people close to them.

Archival photographs can be selected from photo libraries, professional associations, organizational archives, visual data banks, online stock images, and private archives (Ray & Smith, 2012). Such images are chosen to represent an element of the main topic of the research, and to encourage association and interpretation on this topic. For example, an image may be related to an historical event, one that can

> ...trigger interviewee's identification with the activities or phenomena represented and produce reminiscences about the general nature of work, beliefs, and practices in the particular location or industry during that historical period (Parker, 2009, p. 1117).

Archival photographs produce interesting interpretations because they are as much about the present as they are the past. They stimulate here and now reflection on the current state, and they mobilize personal perceptions and projections, helping the respondent to position him or herself emotionally within the social or organizational issues being investigated (Kjellstrand & Vince, 2017; Warren, 2009).

Finally, researchers can utilize a combination of photographs produced by different sources (Ray & Smith, 2012). Researchers and research participants can jointly produce images of the same place prior to photo-elicitation interviews. For example, when analyzing photographs taken by a research group and a group of young prisoners on remand, Sievers (2013) noticed the paradoxical nature of what people show with their photos. According to this study, the photo taken by the researchers "showed a 'pervasive beauty', an esthetic view of the uncanny, ugly, and frightening" where the photos taken by the prisoners "were often reflective of the ugliness of the place, its incarnated violence, and were, thus, an expression of their hopelessness, lament, revolt, and despair" (Sievers, 2013, p. 139). In this study, the researchers and the research respondents each had their own purpose and own agenda, offering opportunities to stimulate dialogue.

Photographs as Data and Photographs as a Device to Generate Data

Photographs can be the data in the sense that they are derived, and initially analyzed, by a respondent or respondents. For example, images taken by respondents can depict objects of interest to the research, which then form the basis for thematic analysis (Warren, 2005). Photographs serve as a cue to remember certain stories that otherwise might have been forgotten. In this way,

> ...photographs, along with their layers of historical meaning, implicit in image makers' intension, subjects' representations and viewers' interpretations, can play an important role in teasing out and the sense making of organizational complexity (Parker, 2009, p. 1114),

Especially when the research question concerns the perception of participants (Tyson, 2009). Photographs can also be used as a device to stimulate broader conversations. For example, in their study of hair salon workers, Shortt and Warren (2017) analyze the meaning that photographs have for both the researchers and respondents who have taken photos; and they consider the broader field or sample level meanings interpreted from analysis of "image-sets."

Thus, photographs provide researchers and respondents with opportunities for in-depth reflection, creating an object that generates a "third-party effect" (Warren & Parker, 2009). The inclusion of photographs during interviews shifts the emphasis from one-to-one interaction between researcher and respondent and can decrease the power distance between them (Ray & Smith, 2012).

PROJECTIVE PHOTO-ELICITATION

The research that we discuss in this chapter used photographs as objects on which respondents could reflect. We introduced a device that encouraged respondents to *project* their feelings into two images that were chosen to represent two sides of peoples' lived experience of different social systems. Our aim was to capture elements of the social and political context in which respondents were embedded, but to do this through their emotional investments in contrasting images. We found that respondents had very strong reactions to the contrasting images of Soviet and post-Soviet life. The images stimulated personalized experiences and preferences, tensions that were inherent in the transition from the old to the new system, and broader social and organizational dynamics relating to collective experience of these tensions. We discuss these in detail below.

Organizations are "driven, inhibited and guided by different emotions, including fear and hope, excitement and despair, curiosity and anxiety..." (Antonacopoulou & Gabriel, 2001, p. 444). Scholars have used photographs to evoke emotions, and thereby to identify social emotions (emotions that tie communities together, see Creed, Hudson, Okhuysen, & Smith-Crowe, 2014); as well as individual and social defenses against emotion that limit organizational members' ability to act (Sievers, 2013; Warren, 2012, 2002). Our interest was also in the difference between emotional responses made in the interviews and in relation to the photographs, and our empirical design provided a unique opportunity to access this dynamic.

Indeed, one of the clearest aspects of our research was that there is a profound difference between life in the old Soviet and the new free market systems. However, we did not imagine that changes in socioeconomic design had made a difference to the Soviet mentality in Kazakhstan, or that it was possible to eliminate Soviet era perceptions from free market practices (Kjellstrand & Vince, 2017). Our interest was rather in the tensions mobilized between the "old" and "new" in order to comprehend the ways in which they combined both for people and for organizations. In addition to peoples' experience of the tensions implicit in the process of transforming from one system to another, we sought to identify aspects of the (hybrid) dominant order that were reflected in social and organizational dynamics. How respondents interacted with colleagues emerged from projections onto the (imagined) relationships represented in our chosen images. Our interest, therefore, was to capture collective dynamics through respondents' association with the "others" in the different images. By using them in this way, photographs can bridge "the gap between the apparently individual, private, subjective and the apparently collective, social, political" (Samuels, 1993, p. 63).

In using this method, we were seeking to explore three interrelated questions. First, what differences in expression of emotion could be identified by comparing our interview data with our photo-elicitation data? Second, how does photo-elicitation help us to comprehend different combinations of the "old" and "new" systems? Finally, how does photo-elicitation capture collective dynamics through respondents' associations?

THE EMPIRICAL STUDY

The information gathered for this study is part of a larger empirical data set collected in 2014 at five organizations located in Kazakhstan. This is a country undergoing transformation from a Soviet system with a planned economy toward a capitalist-free market economy. We conducted 52 semistructured interviews for the research. At the end of each interview, we introduced a pair of photographs, one depicting an old Soviet workplace and the other a contemporary workplace (see photos 1 and 2, below). Our aim was to contrast the two sets of stimuli and encourage the participants to free associate their views of the old and the new. The two photographs were introduced to respondents at the same time and they were asked about what they felt while looking at them. Each participant spent around 10–15 minutes looking, talking, and comparing these two photos, which resulted in 80 pages of transcribed data.

The interviews were conducted in either the Kazakh or Russian languages, depending on the preference of the participant. Twelve interviews were directly translated and transcribed by the first author, and the remaining 40 were professionally transcribed, being later coded in their original languages. Any part of these that are cited in this chapter were translated into English at a later stage. Our coding process was inductive and emergent (Gioia, Corley, & Hamilton, 2013). For our initial process of open coding, we were specifically interested in capturing the various emotions people felt and expressed while looking at the photos. In our second round of coding, we categorized similar emotions into themes.

Description of Photos We Used during Our Photo-elicitation

Two photos were introduced to the participants at the end of each interview. The first (photo #1) was used to represent the Soviet workplace, specifically a teacher's room at a school. This photograph was sourced from a private archive. We offered it with the aim of showing the Soviet workplace, which represented the old times from which the organizations have been moving away.

The second photo is used to depict a contemporary workplace (photo #2), namely an afternoon staff meeting. The photo was sourced from the Wikimedia Commons website, which offers unlicensed photos for public use. The main purpose of showing this photo was to offer participants an image of the contemporary market economy workplace.

Photo #1. A Soviet Workplace – From a Private Archive.

Photo #2. Afternoon Staff Meeting – From Wikimedia Commons by Robert Scoble.

FINDINGS

Excerpts from three interviews are presented in this section to illustrate how photo-elicitation can bring out emotions, tensions, and group dynamics. Although we conducted separate analyses for each method, the data we worked with include dynamics between people, and carry narratives that are complex and difficult to easily divide and separate following the theoretical concepts. In other words, data have been segmented here solely to simplify and organize our findings.

Vignette 1: Emotions Projected onto the Photographs

This interview proceeded in a positive tone and the participant, a camera operator, was talkative when discussing workplace relations and the details of his job. Marat (pseudonym) went into great technical details when describing his own work, which involved making short films, videos, and photography. The participant described mostly positive work relations. According to him, workplace relations revolved around people helping each other, both within and outside work. In his words:

> We all work together. For instance, if I do not work, maybe work will not stop, maybe it stops a bit. But if our journalists do not work, then everything stops. I am telling you, journalists – they are the newspaper. They work, we help – they are the breadwinners... we facilitate their work – everything we can do for them; we help them with the photos. If necessary, I go and take a new photo or find something suitable from the archival photos or videos. If we have only old videos, and they need a photo on that topic, I do 'stop cadre' and take a photo of it for them; anything that is needed – I will get it for them! It is our common goal. (Marat, a camera operator)

In the standard interview, the respondent conveys a strong sense that work relations at his organization evolve around people helping each other. Marat values his journalist colleagues and appreciates their hard work. He describes himself as a supportive person, who facilitates their work and helps with everything the journalists may need. He feels personally responsible for providing photographs for the newspaper and he feels confident he can produce the required photo whatever it costs. It feels like he goes the extra mile to help his colleagues. When he speaks of this, his voice is calm, and he communicates a sense of solidarity. He is happy and proud to be part of the collective.

However, when the pair of photographs was introduced (photos #1 and #2), the participant's emotional state changed. Strong and opposing feelings emerged as he looked at the images:

> You cannot compare these two photos, if you are asking ... This photo... (photo 2) here... I hate them, these ones... (he speaks louder) None of them can do the job! Look at the way he sits, here, you can see immediately... One leg on the other, hands are like this... (loathsome tone of voice, speaks fast and with scorn). One should be modest, like there (showing the black-and-white photo; voice is calm.) Like them. These guys, they really worked hard. Those were the days when people really worked hard. (Back to photo 2) What is this?! (voice speaks louder) As if he sits at home, isn't this an office?! I noticed that immediately, from the second they entered the office, smoking... smoking in the office that is too much... (Marat, a camera operator)

Marat suggests that the people depicted in photo 2 cannot do their job. The participant says that the way that the people are sitting in the photo is wrong. He is angry toward those who do not do the job at work. The tone of his voice is high, the way he describes the photo is intense, and he is scornful when speaking about this photo. He compares the people he is angry with in photo 2 with those in photo 1 who are seen as modest and hard working. When he describes photo 1, his voice is calm and proud. His reaction to photo 2 indicated this participant's dislike of the new work order and his preference for the old one. His projections onto the images allowed him to *locate* himself on one side in a way that was not apparent within the interview. This was particularly represented by the emotional intensity of his reaction to the photos, in contrast to the positive tone of the interview. The photograph thus elicited data beyond the data generated by the interview alone by generating intense and revelatory emotions.

Vignette 2: Tensions Projected onto the Photographs

Our second example is from an interview conducted with a deputy headmistress, Aisha, at a school situated in a small village. Aisha was an open and talkative person, who eagerly shared her views on workplace relations and the socially shared rules there. She described her community as caring, where colleagues supported each other in any situation. She gave an example, when a colleague needed a new placard/poster and they did not have any budget for it. "All colleagues gathered a bit of money each and ordered the placard from the nearest city via a taxi." She also explained that the community could be divided into young, middle-aged, and older teachers. While the young teachers were those who had worked there only for one or two years, the older ones were those who had a couple of years left until retiring. The middle-aged teachers had been working there for a substantial amount of time and made up the bulk of the group. A story offered by the respondent as an example of tensions at work during the interview concerns the scheduling at work and the attitude of teachers:

> Older teachers... they seem to react to any decision I make. They seem to get offended easily, I think, and they complain. I made the schedule for the next semester the other day. I tried to accommodate all the teachers, take into consideration their needs. For instance, younger teachers have kids, they need to cook and take care of their kids when they come home. So, I put a couple of gap hours in the teaching schedule for the older teachers. [It seems teachers prefer to teach all their classes one after another and go home, instead of having gaps in between and have longer working days]. So, they complained, as I heard afterwards. "Isn't it better with a little bit of a time in between your classes? So, they could prepare for the next class instead of doing it at home." No, they do not want gaps between their classes! Such things create conflict between us. But I try to keep silent and not to talk too much; what can I do? (Aisha, Deputy Headmistress)

The participant justifies her reasons behind the scheduling by saying that she is trying to take each teacher's family situation into consideration. Aisha scheduled the classes of younger teachers consecutively and those of older teachers with gaps. Even though she tried her best, she could not satisfy everyone, and those who had gap hours between their classes complained. She wanted to avoid conflict with the teachers, and she was ready to remain silent to avoid further escalation. She also highlighted the importance of keeping positive relations between all the teachers and mentioned that, "one cannot shout and order them about," because "I want to be on good terms with all of them."

After the photos were introduced, she recognized that they depicted the old and new systems, and she described her views on the two. Unlike in vignette #1, she did not display any strong emotions toward the photos. Rather, she mentioned the technical differences between the old Soviet and the new times, with photo 2 leading her to talk about the technological advancements in teaching and various new possibilities new technology brought to the profession. When Aisha was asked about the differences she saw in the photos, she did not contrast them, but rather, explained how they seemed to complement each other. She raised some issues about the older and younger teachers, projecting what she thought of the former onto photo 1 and her thoughts regarding younger teachers onto photo 2:

> Everybody is by themselves. They are scowling. They seem to be a bit selfish and callous. Everybody is by himself or herself… I think so… yes… the young (teachers) do not want to say what they know. The older (teachers) do not want to write the papers. Maybe they do not have time sometimes… here the young teachers seem greedy and do not care about others (pointing at photo 2). We (middle aged teachers?) give away what we have in our hands, but the young teachers are tough, they do not give anything away. (Aisha, Deputy Headmistress)

This participant does not consider the photos separately. She associates them with the tensions between the older and younger teachers in the school. She feels that the people in both of the photos are "by himself or herself." This had a negative tone when spoken. In the interview, she glossed over the tensions within the school and emphasized what a strong community it is. She was able to delve deeper into these tensions using the photos. The people in the images became her own colleagues (both young and old) and she used the photo to communicate that they appear to be selfish and callous. The young teachers do not want to say what they know, and the older teachers do not do what they are asked to do. In the interview, while she divided her colleagues into three groups, overall, she focused on the positive, emphasizing her wanting "to be on good terms with all of them." During the photo-elicitation, it was possible to also express and delve into the negative thoughts she had about them. She also mentioned in the photo-elicitation that "we give away what we have in our hands," referring to the group of teachers she belongs to as being generous, as compared to the young ones, who seem to be more selfish. This does not align with her engrained value of teaching involving generosity toward colleagues. The older teachers are also resistant to her view of what they should be doing. Neither group complies with her sense of the social rules, and therefore the imagined cohesiveness of the community fades in the second part of the interaction, supported by photo-elicitation.

Vignette 3: Social and Organizational Dynamics Projected onto the Photographs

In an interview with Maya, a teacher, the respondent was positive about her workplace. When she was asked whether her workplace was any different from other similar workplaces, she said that it was much more interesting to work here. "I would not go elsewhere even if they invited me." Maya also said that "being a teacher means a lot of responsibility and there are other positive aspects to being a teacher as well." She seemed to like being a teacher and being in her particular workplace. She spoke about working together and that people were polite and respectful of each other.

> There are certain social rules and regulations in the collective... generally the relationship among colleagues is not bad at all... People do not speak bad of each other... Nobody says "you are wrong"; people just politely mention if there is anything that needs to be said. Even if there are issues among members, we try to solve the problems right there without making a big deal out of it. People talk to each other; they explain what is what. (Maya, Teacher)

During the interview, Maya described the relations between collective members as positive. People are agreeable and try to see things in a positive light. They are open with each other and would aim to solve any problem together, "right there without making a big deal of it." Her perspective changed when the photographs were introduced. The images sparked a reflection about one particular staff member, the school psychologist:

> Between you and me... I do not like the way our current psychologist works. I even mentioned that in the previous teachers' meeting. She gives questionnaires to the students, and if they get any negative response about someone, they come and blame us (teachers)... if someone gets F; it is the teacher's fault... Then, "there was a better teacher last year, this teacher is not good enough" is the story among parents. I wrote about it in the questionnaires, but nobody acts upon our feedback. If only we could work together, think together... that is our problem here... I cannot say what I think; if I say that, they [other teachers] will talk about it in their own circles. They do not talk openly and try to solve the problem; instead, they blame the person who pinpointed the issue. Then, I hear about it from someone else. They upset me, I upset them; we upset each other. (Maya, Teacher)

Maya resented the ways in which the school psychologist used questionnaire results, mentioning that it affected the way parents reacted to them. She regretted that nobody seemed to respond to her own feedback on the questionnaires. Things had escalated, with some of her colleagues taking the side of the psychologist, and some parents blaming her for the students' poor grades. Teachers were divided, discussing the problem separately with like-minded colleagues without openly talking about it. Discussing it behind each other's backs, they became caught up in blame. She explained how working together seemed to be a problem for her, because the workplace was unable to find a solution together. It seemed to her that the person pinpointing a problem was likely to get the blame for it. While in the interview she described her school as an interesting place to work, in the photo-elicitation, she started to delve into greater depth, revealing other dynamics.

DISCUSSION AND CONCLUSION

Our purpose in this chapter is to discuss the potential of photo-elicitation as a data generating device. We have used our experience from an empirical research project to show how semistructured interviews and photo-elicitation worked together to produce richer data. In particular, we found that these methods brought out respondents' emotions in different ways, and that this made for interesting comparisons between the findings from these two qualitative approaches. We were struck by the intensity of emotion and strong associations that emerged from respondents' engagement with the photographs, and the difference from their response to interview questions, which tended to be more controlled and positive.

We make a methodological contribution by showing how to use photos to enrich qualitative research designs based on interviews. In our experience, the use of photographs at the end of the interview produced data with more emotional intensity and directness from respondents. The emotional conflicts that they felt in relation to their coworkers or toward their organizations seemed to us to be a more honest reflection of tensions within the Kazakhstan social system. We would argue that (within our research context) photo-elicitation was most effective in drawing out peoples' lived experience of the social context being investigated. Having reflected on our experience, we think that there were two main advantages of using photo-elicitation in combination with semistructured interviews in our study.

First, respondents expressed their emotions more freely during the photo-elicitation. The interviews tended to be conversational, affording respondents the opportunity to offer thoughts, opinions, and narratives of their experience. They were professionals, who were fluent in their distinct professional discourses. Of course, these conversations were not free from emotion. However, when the photographs were introduced, we found that most of our respondents become more emotionally charged and less constrained in their responses. The images had the effect of delving deeper into participants' responses because they so vividly represented within the social system (changing from a Soviet to a post-soviet economy). This helped us to show how embedded these structural tensions are in the lived experience of professional workers.

The photographs brought out negative and complex emotions, which were generally absent or glossed over during interview conversations. Respondents associated the photographs with their own experience, taking them into the space between their personal emotions and their professional role. This space has been called "the organization in the mind" (Armstrong, 2005; Kjellstrand & Vince, 2017). This refers to the ways in which individuals respond to their experience within organizations, how implicit rules and expectations become personalized/internalized, and therefore to the emotions that tie people to organizations in similar and different ways. For example, in an interview it can be difficult to uncover someone's dislike toward a colleague or job situation. Our respondents tended to be positive in their answers when asked about work dynamics, even when they were sharing negative experiences. Using photo-elicitation allowed participants to speak what was actually in their minds. A comparison between words used in the interview and the photo-elicitation illustrates this change: "they complain" becomes "they are scowling"; "we facilitate their work" becomes "I hate them!"; and "we try to solve the problems right there" becomes "I cannot say what I think."

Second, as we studied a system in the process of change (in between the "old" and the "new"), photo-elicitation helped to comprehend different combinations of peoples' lived experience of the old and the new that coexisted within what is, in effect, a hybrid or dual system. Associating the two images uncovered ongoing tensions entwined with the process of change. For example, one respondent compared the two photographs and highlighted the technical differences between them. One image was associated with the pen-and-paper period and the other with the "new" more technologically advanced times. The respondent expressed her wish to have better technical equipment, to be able to rise up to the expected standards of the new system. Another respondent made a similar association and added that "there is a TV in the photo – if we had such a TV and (we could) hold some interesting activities, our motivation would improve." As with other professional areas, the school system in Kazakhstan is undergoing considerable reform requiring fundamental changes. Her narrative tells us something specific about the current state of systemic changes and about her simultaneous experience of the "old" and "new" times.

While these two points reflect our experience of the difference between semi-structured interviews and photo-elicitation in our research, we also believe that such designs can be more generally useful in management and organization studies. Ray and Smith (2012) have highlighted the scarcity of photographic research methods that explore processes in organizations. We think that photo-elicitation is a useful method for researchers seeking to capture established and emerging dynamics and processes through respondents' associations, alongside their personal perceptions, ideas, and assumptions. Respondents internalize the photographs and interpret them by associating their actual and imagined content with their own stories (Warren, 2005). These stories frequently involve the groups to which people belong, the relationships between people, and the social context within which these relationships take place. Simply put, respondents both perceive the organization in the images and feel their experience of it. Our view is that photo-elicitation is therefore potentially important in the development of themes and issues pertaining to intrapersonal and interpersonal aspects of organizational processes as people engage with the possibilities and impossibilities of change.

We think that there are particular areas of management and organization studies that could benefit from the integration of photo-elicitation into an overall qualitative research design. The most obvious of these relates to the study of emotion in organization and to the relationship between emotion and reason. A design based on interviews alone is more than capable of generating insights into individual and collective emotion, and the rational structures that contain them (see Vince, 2006 for an example). However, our research suggests that photo-elicitation is helpful in producing data at the interface of emotion and *structure*. This makes it particularly relevant as a method for the study of social emotions, which are "emotions that pertain to the state of the social relations... that hold communities together in institutional processes" (Creed et al., 2014, p. 276). We think that photo-elicitation will be useful to researchers with an interest in studying the interplay between processes of organization and subjectification.

To conclude, we have had an experience of using photo-elicitation that has made us enthusiastic about recommending this method, especially where it is set alongside other qualitative methods. We think that the approach offers the potential for increased access to the depth or intensity of the emotional experience being studied. We have particularly emphasized the value of the *projective* qualities of the images in our context. The chosen images mobilized very powerful associations for respondents and provided us with insights about personal and social emotions (fear, anger, disappointment) embedded within peoples' attempts to move from one system to another. Photo-elicitation is a device that researchers can utilize in the service of comparison between participant responses within interviews and the underlying emotions that both infuse and confuse their experiences. We encourage other researchers to explore the generative potential of photo-elicitation methods.

REFERENCES

Antonacopoulou, E. P., & Gabriel, Y. (2001). Emotion, learning and organizational change: Towards an integration of psychoanalytic and other perspectives. *Journal of Organizational Change Management*, *14*(5), 435–451.

Armstrong, D. (2005). *Organization in the mind*. London: Karnac.

Boxenbaum, E., Jones, C., Meyer, R. E., & Svejenova, S. (2018). Towards an articulation of the material and visual turn in organization studies. *Organization Studies*, *39*(5–6), 597–616.

Creed, W. E. D., Hudson, B. A., Okhuysen, G. A., & Smith-Crowe, K. (2014). Swimming in a sea of shame: Incorporating emotion into explanations of institutional reproduction and change. *Academy of Management Review*, *39*, 275–301.

Davison, J., McLean, C., & Warren, S. (2012). Exploring the visual in organizations and management. *Qualitative Research in Organizations and Management*, *7*(1), 5–15.

Davison, J., McLean, C., & Warren, S. (2015). Looking back: Ten years of visual qualitative research. *Qualitative Research in Organizations and Management: An International Journal*, *10*(4), 355–359.

Gioia, D. A., Corley, K. G., & Hamilton, A. L. (2013). Seeking qualitative rigor in inductive research: Notes on the Gioia methodology. *Organizational Research Methods*, *16*(1), 15–31.

Harper, D. (2002). Talking about pictures: A case for photo elicitation. *Visual Studies*, *17*(1), 13–26.

Heisley, D. D., & Levy, S. J. (1991). Autodriving: A photo-elicitation technique. *Journal of Consumer Research*, *18*(3), 257–272.

Hurworth, R., Clark, E., Martin, J., & Thomsen, S. (2005). The use of photo-interviewing: three examples from health evaluation and research. *Evaluation Journal of Australasia*, *4*(1–2), 52–62.

Jarrett, M., & Liu, F. (2018). "Zooming with": A participatory approach to video ethnography in organizational studies. *Organizational Research Methods*, *21*(2), 366–385.

Kjellstrand, I., & Vince, R. (2017). No room for mistakes: The impact of the social unconscious on organizational learning in Kazakhstan. *Administrative Sciences*, *7*(3), 27.

Meyer, R. E., Höllerer, M. A., Jancsary, D., & Van Leeuwen, T. (2013). The visual dimension in organizing, organization, and organization research: Core ideas, current developments, and promising avenues. *The Academy of Management Annals*, *7*(1), 489–555.

Padgett, D. K., Smith, B. T., Derejko, K. S., Henwood, B. F., & Tiderington, E. (2013). A picture is worth…? Photo elicitation interviewing with formerly homeless adults. *Qualitative Health Research*, *23*(11), 1435–1444.

Parker, L. D. (2009). Photo-elicitation: An ethno-historical accounting and management research prospect. *Accounting, Auditing & Accountability Journal*, *22*(7), 1111–1129.

Petersen, N. J., & Østergaard, S. (2003). Organisational photography as a research method: What, how and why. In *Academy of management conference proceedings*. Submission Identification Number 12702, Research Methods Division. Retrieved from http://citeseerx.ist.psu.edu/viewdoc/download?doi=10.1.1.561.250&rep=rep1&type=pdf

Radley, A., & Taylor, D. (2003). Images of recovery: A photo-elicitation study on the hospital ward. *Qualitative Health Research*, *13*(1), 77–99.

Ray, J. L., & Smith, A. D. (2011). Worth a thousand words: Photographs as a novel methodological tool in strategic management. In D. J. J. Ketchen (Ed.), *Building methodological bridges* (pp. 289–326). Bingley: Emerald Group Publishing Limited. doi:10.1108/S1479-8387-0000006013

Ray, J. L., & Smith, A. D. (2012). Using photographs to research organizations: Evidence, considerations, and application in a field study. *Organizational Research Methods*, *15*(2), 288–315.

Samuels, A. (1993). *The political psyche*. London: Routledge.

Shortt, H. L., & Warren, S. K. (2017). Grounded visual pattern analysis: Photographs in organizational field studies. *Organizational Research Methods*, *22*(2), 539–563.

Sievers, B. (2008). 'Perhaps it is the role of pictures to get in contact with the uncanny': The social photo matrix as a method to promote the understanding of the unconscious in organizations. *Organisational and Social Dynamics*, *8*(2), 234–254.

Sievers, B. (2013). Thinking organisations through photographs: The social photo- matrix as a method for understanding organizations in depth. In S. Long (Ed.), *Socioanalytic methods: Discovering the hidden in organisations and social systems* (pp. 129–151). London: Karnac.

Slutskaya, N., Simpson, R., Hughes, J., Simpson, A., & Uygur, S. (2016). Masculinity and class in the context of dirty work. *Gender, Work and Organization*, *23*(2), 165–182.

Stamenova, K., & Hinshelwood, R. D. (Eds.). (2018). *Methods of research into the unconscious: Applying psychoanalytic ideas to social science*. Abingdon: Routledge.

Tyson, T. (2009). Discussion of photo-elicitation: An ethno-historical accounting and management research prospect. *Accounting, Auditing & Accountability Journal*, *22*(7), 1130–1141.

Vince, R., & Warren, S. (2012). Participatory visual methods. In *Qualitative organizational research: Core methods and current challenges* (pp. 275–295). London: Sage.

Vince, R. (2006). Being Taken Over: Managers' emotions and rationalizations during a company takeover, *Journal of Management Studies* *43*(2), 343–365.

Warren, S., & Parker, L. (2009). Bean counters or bright young things? Towards the visual study of identity construction among professional accountants. *Qualitative Research in Accounting and Management*, *6*(4), 205–223.

Warren, S. (2002). "Show me how it feels to work here": Using photography to research organizational aesthetics. *Theory and Politics in Organizations*, *2*, 224–245.

Warren, S. (2005). Photography and voice in critical qualitative management research. Accounting, *Auditing & Accountability Journal*, *18*(6), 861–882.

Warren, S. (2009). Performance, emotion and photographic histories: A commentary on professor lee D. Parker's paper. Accounting, *Auditing & Accountability Journal*, *22*(7), 1142–1146.

Warren, S. (2012). Psychoanalysis, collective viewing and the "social photo matrix" in organizational research. *Qualitative Research in Organizations and Management: An International Journal*, *7*(1), 86–104.

CAPTURING ORGANIZATIONAL COMPASSION THROUGH PHOTOGRAPHIC METHODS

Timothy M. Madden, Laura T. Madden, and Anne D. Smith

ABSTRACT

This chapter highlights the value offered by photographic research methods to the study of organizational compassion. We demonstrate this potential by first briefly reviewing the history and usage of photographic research methods in the social sciences and the state of compassion research. We then describe how compassion emerged as a key theme in a field study that utilized photographic methods. From this, we identify four approaches that photographic research methods can be used to extend our understanding of compassion in organizations. Specifically, we clarify how this stream of research can be enhanced by the inclusion of photographic methods. We highlight critical research decisions and possible concerns in implementing photographic methods. The chapter concludes with additional organizational phenomena that would benefit from using a photographic methods approach.

The various methods gathered under the umbrella label of qualitative (Guba & Lincoln, 1994), defined as the study of "things in their natural settings, attempting to make sense of, or interpret, phenomena in terms of the meanings people bring to them" (Denzin & Lincoln, 2005, p. 3), offer many benefits through their ability to access, explore, and experience real organizational people and problems in rich detail (Van Maanen, 1979). As an example, photographic research methods—primarily qualitative methods through which researchers use photographs to elicit information during interviews and focus groups—often result in deep and nuanced data (Collier & Collier, 1986; Harper, 2005; Vince & Warren, 2012). Photographic

Advancing Methodological Thought and Practice
Research Methodology in Strategy and Management, Volume 12, 55–71

ISSN: 1479-8387/doi:10.1108/S1479-838720200000012016

methodologies are well-suited to the exploration of new phenomena because they allow researchers to get close to the lived experience and organizational processes (Dion, 2007), attend simultaneously to the social and material world in organizations (Shortt & Warren, 2012), and offer the potential to "mine deeper shafts into a different part of human consciousness than do words-alone interviews" (Harper, 2002, p. 23). Organizational research has traditionally been dominated by a positivistic paradigm that focuses on theory evaluation through the use of quantitative methodologies (Lin, 1998; Sutton, 1997), whereas qualitative research offers the potential to build theory by illuminating underlying processes and causal mechanisms in specific contexts (Lee, 1999). Researchers developing theory may be particularly interested in the richness of the data gathered with qualitative methods (Edmondson & McManus, 2007) such as photographic methods. Qualitative research is thus well-matched to nascent literatures that require inductive study about a phenomenon to generate foundational knowledge (Edmondson & McManus, 2007).

One such nascent research stream that could benefit from photographic methodologies is organizational compassion (Rynes, Bartunek, Dutton, & Margolis, 2012). In its current state, compassion research within the organizational literature has generated many narratives of experiences of compassion in response to a specific tragedy (Dutton, Worline, Frost, & Lilius, 2006), as an organizational capability (Lilius et al., 2011b), or as an organizational capacity that an organization can develop (Madden, Duchon, Madden, & Plowman, 2012). These stories demonstrate that the common elements of the compassion process are the noticing of someone else's pain, empathizing with that person, and then responding in a way designed to lessen that pain (Kanov et al., 2004); however, because this process is so individualized, photographic methodologies offer researchers a chance to capture valuable new information about this process and the experience of compassion within organizations. In this chapter, we describe many potential benefits of designing organizational compassion research based on photographic methodologies.

In doing so, we offer several contributions. First, we show how photographic methodologies can create deeper responses during interviews and observations that may lead to surprising insights for theory. Second, by suggesting some of the insights that have been generated about compassion through photographic methodologies, we offer novel research ideas for this growing body of literature. The following sections provide background on the development and history of photographic methodologies and review the studies and methodologies that have contributed to our understanding of compassion within organizations. Subsequently, we describe some of the ways in which compassion has surfaced during our own field study using photograph elicitation. Finally, we describe possible studies that could benefit from the use of four forms of photographic methodologies to explore more targeted research questions related to organizational compassion and also offer a range of other organizational phenomena that could benefit from a photographic methods approach.

Keywords: Qualitative methods; research methods; photo elicitation; compassion; organizational culture; field study

LITERATURE REVIEW

Photographic Methods in Organizational Research

The use of photographs in social scientific research has a rich history; particularly in anthropology and sociology (Banks, 2007; Harper, 2005; Wagner, 1979). In these disciplines, photographic research methods are used to convey the reality of the field setting and as a means to invoke a visceral reaction in the audience in a way that text alone cannot (Harper, 2002). In anthropology, photographic methods have continued to be an intrinsic part of in-depth ethnographic studies. Bateson and Mead (1942) used photographic methods to capture Balinese culture; in one set of photographs, they were able to capture the emotionally charged process of a trance (Stasz, 1979). In sociology, the early use of photographs was primarily to illustrate field settings and to shock and incite people to action by highlighting deplorable living and working conditions (Blackmar, 1897; Bushnell, 1902). The Farm Security Administration documentary in the 1930s captured the pathos of the depression; these photographers had an impact on the later growth in visual sociology (Banks, 2007). This type of field documentation in both sociology and anthropology has been criticized as privileging the researcher's view of reality, as well as the possibility of staging photographs and not allowing participant voices to be heard. Despite these shortcomings, this approach was able to portray a situation's pathos and elicit readers' emotions, both of which are essential to compassion research.

In the 1960s, use of photographs in field research began to shift its focus from readers of published images attached to field studies to photographs that are shared with research participants. Anthropologists Collier and Collier (1986, p. 131) recognized that

> ...photographs are charged with psychological and highly emotional elements and symbols... [but offered that] the only way we can use the full record of the camera is through the projective interpretation by the native.

They introduced photo-elicitation approaches, which included native interpretations of the meaning of photographs, and found a profound richness in the field participant's explanation and reaction to photographs they had taken of their society or family. Reflecting an interpretive turn in sociology, Becker (1974) called for more use of photographic methods and inclusion of the participants' voice. Visual sociology and anthropology's use of photographic methods has grown rapidly since the 1970s (Banks, 2007; Margolis & Pauwels, 2011; Prosser, 1998), but these approaches have not been fully incorporated into studies of management and organizations despite calls for inclusion to understand organizational life (Meyer, 1991).

Photographic methods in management research have faced slow adoption (Ray & Smith, 2012), although, similar to the early studies in sociology, a few journal articles have included them, mostly to illustrate field settings (e.g., Dacin, Munir, & Tracey, 2010). A few exceptional studies have used photographic approaches, such as a study that allowed participants to see and comment on their role in an organizational process (Buchanan, 2001). Another study had participants take photographs of how knowledge is shared in their organizations and then discuss their photos as a group (Petersen & Østergaard, 2004). Other researchers have tried to get at the lived experience of participants, such as individual sensemaking of retail spaces (Venkatraman & Nelson, 2008), but still only a few studies of organizational elements have used photographic methods.

One of the most promising lines of photographic approaches in organizational research is the ability to capture elements of corporate culture by having organizational participants take photographs. For instance, Warren (2002) examined individual feelings in the workplace; she engaged in an ethnographic study of a high tech firm and asked participants to take photographs of "what does it feel like to work here?" Warren (2002, p. 230) argues that

> ...in order to explore the relationship between the feel, sights, smells, and tastes of the organizational setting and the people who work there, surely a more "sensually complete" methodology than a narrow and limiting focus on those aspects of organization which can be spoken or written down is demanded.

Through the use of the tech workers' photographs as the focal point of the semistructured interviews, Warren (2002, p. 242) is able to examine aspects of organizational life such as emotions that would otherwise be difficult to capture, much less discussed, because of their "tacit, intangible, and largely ineffable nature." For these reasons, photographic methods are well-suited to address the growing research interest in the organizationally-relevant social phenomenon of compassion.

Compassion in Organizational Research

At the individual level, compassion is a three-stage social process that involves noticing the suffering of another person, followed by feeling a sense of empathy with that person, and then engaging in a response in an attempt to alleviate the other person's suffering (Clark, 1997). Compassion is described as an innate human response to another's suffering (Frost et al., 2006, pp. 843–866), and this three-stage process, notice–feel–respond, forms the foundation of organizational compassion (Kanov et al., 2004) and how organizational members respond to each other following the personal tragedy of another member (Dutton et al., 2006; Lilius et al., 2011a; Madden et al., 2012). Experiencing compassion at work is positive for organizations, namely, it has been associated with reduced employee turnover (Lilius et al., 2011b) and deeper affective commitment (Lilius et al., 2008). Additionally, employees have been found to be attracted to and remain in particularly compassionate work units (Frost, Dutton, Worline, & Wilson, 2000) even if the organization itself is not seen as equally compassionate.

At the organizational level, compassion is a collective process through which individual efforts coalesce into organization-wide expressions of compassion

(Kanov et al., 2004). Examples of organizational compassion from this research include a university-wide response to students affected by an apartment fire (Dutton et al., 2006), company responses to organizational members and their families following a terrorist attack (Dutton, Frost, Worline, Lilius, & Kanov, 2002), and compassionate responses from a work unit that had institutionalized compassion and care for its members as a core value (Lilius et al., 2011b). Organizational compassion has been examined as the product of endogenous resourcing (Dutton et al., 2006), everyday routines and practices (Lilius et al., 2011b), and emergence (Madden et al., 2012). With repeated use, compassion becomes a reliable capability of the organization (Lilius et al., 2011b) to be employed as a buffer in the face of less-severe instances of suffering (Cameron, Bright, & Caza, 2004).

Despite this surge in interest over the past 10 years, organizational compassion still meets several criteria for description as a nascent research area (Edmondson & McManus, 2007). The majority of the studies within this stream have been qualitative in nature, with open-ended research questions, such as "What are the foundations of this unit's compassion capability?" (Lilius et al., 2011b, p. 878). Most data gathering has occurred through narrative analysis (Lilius et al., 2008) or interviews and documents obtained through site visits (Dutton et al., 2002; Frost et al., 2000), and constructs are typically new, with few or no formal measures (Lilius et al., 2011b), and often described as "exploratory" (Lilius et al., 2008, p. 213; Lilius et al., 2011b, p. 877).

As research streams mature, with better-defined constructs and measures, research questions become more directed at testing specific hypotheses through the use of quantitative data and validated measures or scales in an attempt to generalize the findings beyond the sample (Edmondson & McManus, 2007). Compassion research is not yet at this stage because many large questions still require qualitative methods. Photo elicitation allows specific participants from specific contexts to provide key insights into their thought processes, their organization, and how they experience the presence (Dutton et al., 2002; Kanov et al., 2004) or absence (Frost et al., 2000; Lilius et al., 2008) of compassion in their daily lives. Because researching compassion may involve relating intensely personal experiences of pain and suffering (Frost et al., 2000; Lilius et al., 2011b), researchers are less likely to want to ask direct interview questions as a way of relating to participants. Therefore, photo elicitation, in which photographs are the central focus of the field interview, is well-matched to study organizational compassion. Although participants may feel uncomfortable responding to direct questions from an interviewer, photo elicitation may generate deeper insights than interviews alone. We suggest that before compassion researchers jump directly from narrative analysis to formal hypothesis testing, novel qualitative methods such as photo elicitation may contribute to our understanding of how compassion develops and is expressed within organizations.

Finding Compassion in a Field Study

Organizational compassion research has yet to incorporate photographs, but photographic approaches would complement field research, which we discovered serendipitously during an in-depth field study. One author investigated the culture of a medium-sized manufacturing firm that primarily designs and fabricates postoperative surgical garments. The intent of this research was to understand the strategy practices to manage a rapidly growing firm. An important finding of the research was that the workers articulated acts of compassion that were closely linked with their willingness to stay at the firm. The resulting low turnover rate allows this firm to manage rapid growth with a clear, well-understood strategic vision and efficient execution.

This field research used interviews with more than half of the 80 organizational members and photographs of the organizational workplace over three years. Originally, the researcher had planned to take photographs of the organization and conduct photo-elicitation interviews with the participants, in an approach similar to Collier and Collier (1986). It was soon clear that the researcher-generated photographs were not meaningful to the organizational members; instead, the researcher listened during interviews to participants' stories and, in response, took photographs of organizational elements they pointed out as important. Each sewing operator who was interviewed had a story of compassion about the organization and its founders, including how the founder had reached out to workers during a difficult time to help them out by paying for a plane ticket or lawyer fees for a family member to immigrate to the United States. Many organizational members stated that if anything happened to them, they knew that the founders would reach out and take care of their families.

Compassion can be seen in this organization in the relationships between the founders and their employees as well as in stories about how the organization reached out to the community. Even before the firm was profitable, it formed a foundation to provide resources to needy families in the community. Funds provided by this foundation helped with the relocation of Bosnian refugees. Furthermore, the firm highlighted its support of several community organizations with artifacts such as thank you letters placed in a central place in the plant (see Exhibit 1). A clerk who had worked for the firm for over a decade shared how the foundation allowed her to touch the lives of others:

> I have gotten the biggest thrill out of helping people I know, taking [them] shopping. ...We've gotten groceries, awesome things, and ... still do it, every Christmas, last year I helped a family, I know this family, he had lost his job and she lost her job ... cut in pay, barely managing, making it, and I got the oldest son ...[he is a] good student football player and all he wanted a new pair of boots, and I got him a pair of boots – my heart does a double take every time I see him with the boots... powerful what they [the company founders] have done.

During this interview, the clerk showed the researcher a box with unpackaged garments in it (see Exhibit 2). She explained that these were garments from a fashion show and could not be sold. Instead of disposing them, the company was paying for them to be shipped to a hospital in Africa for burn victims. This aspect of compassion was captured not only in interviews but also in photographs that emerged from the interviews. For instance, one day when the researcher happened to be present, a sewing operator brought in her young nephew who had

Exhibit 1. Reaching out to the Local Community.

Exhibit 2. Reaching out Globally: Box of Garments to Send to Burn Patients in African Hospital.

been burned. The nephew was being measured to have a special garment made to help in his healing and, more importantly, his comfort (see Exhibit 3). This form of everyday compassion was a key aspect of their culture that was tied to low employee turnover; the photographs were able to capture glimpses and symbols

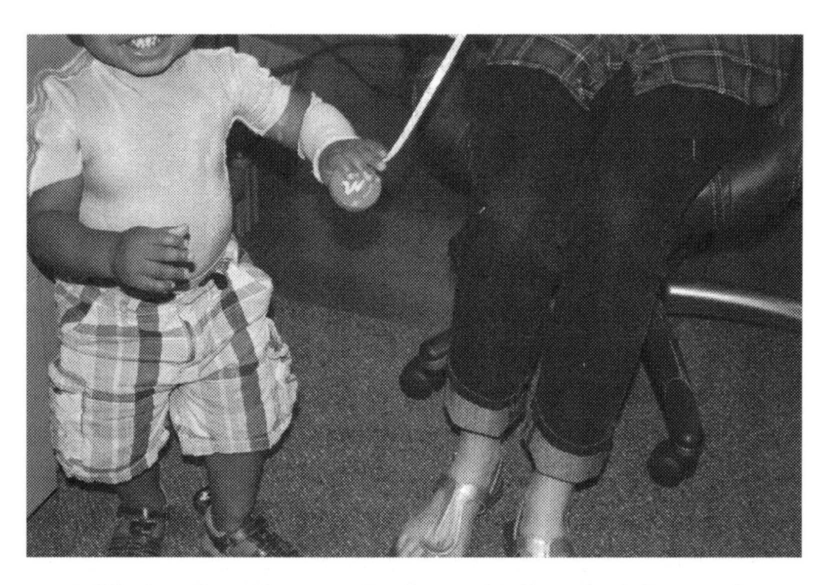

Exhibit 3. Reaching out to Employees: Making Fitted Garment for Employee's Nephew.

of their compassion in a way that provided an enduring reminder to internal and external organizational stakeholders.

This field study revealed that compassion was an important element of this culture that helped in employee retention as well as showing that compassion could be captured in photographs. Below, we suggest ways in photographic methods can be used to research compassion following specific events and as a key value in organizational culture.

STUDYING COMPASSION THROUGH PHOTOS

Based on its potential to contribute to organizational compassion research, we suggest four future studies through which photo elicitation could be used to study compassion. Findings from these studies could benefit researchers and managers interested in the impact of compassion on their organization's culture, improve our understanding of how organizational members respond to acts of compassion around them, and elaborate the benefits that employees perceive from receiving compassion during times of suffering. Table 1 depicts the four studies that could be generated by examining two types of compassion through the two methods of photo elicitation.

Researcher Photograph Study of Compassion as an Event

Proposed Study 1: In order to study a single instance of organizational compassion, researchers could compile archival images related to the timeline of

Table 1. Proposed Compassion Research Studies Using Photographic Methods.

	Researcher Photographs	Participant Photographs
Compassion as event	*Proposed Study 1* Research question: How has this event affected culture and compassion in the workplace? Decisions: Researcher selects archival photos. Individual or group interviews?	*Proposed Study 2* Research question: How has this event affected culture and compassion in the workplace? Decisions: Will participants share their selected archival photos or create new photos? Individual or group interviews?
Compassion as element of culture	*Proposed Study 3* Research question: Is compassion an important element of a workplace? Decisions: Researcher selects own photographs from the organization or use archival or staged photographs that depict some element of compassion (or lack of compassion). Individual or group interviews?	*Proposed Study 4* Research question: How do organizational members understand compassion in their workplace? Decisions: Equipment (disposable camera or cell phones) Statement to guide picture taking (need to pretest). Will the researcher sort pictures or the participant? What will the role of the researcher be (passive time keeper or active moderator) in a group elicitation session?

an event reported by an organization or work unit. Researchers could then interview individuals alone or in focus groups about how the process of compassion unfolded in the face of this event. If the researchers use focus groups, selecting participants at the same level or tenure in the organization to corroborate sequences could enhance the validity of particular interpretations surrounding the incident in question; for example, people new to the organization may have a different understanding of how compassion is related to a particular event. Whether interviewing individuals or groups, the researcher can either take a more involved approach to probe for responses to specific questions or take a more hands-off approach by recording the conversation of group members who talk to each other for a period of time. Another possible variant on this approach would be to have a retrospective on an event some years after its occurrence by placing salient images of compassion on an organizational blog or intranet and allowing organizational members to respond in writing, possibly anonymously, with their reactions to the photos.

One of the shortcomings of this approach is that key images, perhaps organizational members' personal photographs, may be missed when researchers select archival images for the participants to discuss. Another potential hazard of this approach is that it may be emotionally consuming or damaging to either the organizational member or to the researcher, raising a number of potential concerns for human subjects research. Strengths of this approach include investigators' ability to understand how a sequence of compassionate events over the course of weeks, months, or years unfolded as well as a glimpse into how organizational members have made sense of compassion as a facet of their organization's culture. One question to consider with this approach is whether or not compassion has become an institutionalized value following a particular tragedy or if the response was short-lived without any lasting effects beyond addressing the immediate needs of a suffering co-worker.

Participant Photograph Study of Compassion as an Event

Proposed Study 2: The use of participant-provided photographs in response to a single incident of organizational compassion occurs when a researcher invites individuals to provide their own photos of an event or its outcome to the researcher. Participant-provided photographs may come from archival sources such as scrapbooks or photo albums or more contemporary sources such as an assignment where participants are asked to use a camera to document how they believe that this particular incident of organizational compassion has affected the organization. Participants may return from this assignment with images that show a great deal of change or images that show little change despite a major outpouring of compassion for a suffering colleague.

One shortcoming of using participant-provided photographs is that this type of study is likely best suited to individual interviews without a group setting due to the personal nature of these photos, especially if they are owned by the individual(s) who received the organizational compassion. In these cases, the research team may need additional training to handle the emotional impact of this approach. As well, the number of individual interviews will substantially increase the researchers' time in the field as compared to the previously discussed focus group approach. Finally, any researcher analysis of the photographs should be shared with the participants.

Researcher Photograph Study of Compassion as an Element of Organizational Culture

Proposed Study 3: In a project aimed at uncovering rich, interview-based information about compassion as an element of organizational compassion in which researchers create or choose the photographs, researchers would begin with archival photographs or images taken by researchers that depict organizational symbols of compassion in addition to enduring artifacts from prior or ongoing compassionate responses. Additionally, participants might direct researchers toward these images during periods in which the researchers are observing organizational life or conducting informal conversations with participants about particular artifacts or observed behaviors. The images generated for this kind of project could include pictures of cards for a co-worker whose parent recently passed away, boxes used to gather and deliver homemade food for a colleague undergoing chemotherapy, or even thank you notes from the recipients of such compassionate gestures to their officemates. Researchers could take photographs from the work site and then ask participants to explain what those images represent to them. An organization could also gauge impressions of compassion as a facet of organizational culture through an anonymous online photo survey that only identifies the unit of a respondent and presents the respondent with archival photographs of different scenes. Comments on each photograph could be analyzed for internally consistent patterns and comparisons could be made across units to identify different rates of employee turnover, sales growth, or other unit measures.

This process would elicit insights about the presence of compassion in the workplace and may be more likely for participants to respond and provide deeper insights than a battery of survey items. As before, researchers could conduct the elicitation with individual participants or with a focus group of participants from the same work unit or in a group from the same unit. This type of study would allow each organizational member's voice to be heard and enables a comparison across units. Although a focus group may provide more background and meaning about each photo, researchers should be careful to ensure that participants all come from similar organizational backgrounds to ensure that individual responses to the photos are uninhibited; for instance, the power distance that could exist between a CEO and a frontline employee might lead to a focus group session dominated by the CEO's interpretation of the organization's culture, which the frontline employee may be reticent to add to despite having a different viewpoint on the same organization's culture. After conducting a series of interviews with different organizational members, the researchers would compile the interview data for analysis of patterns across and within units of an organization. Researchers could also to compile these findings into an executive summary to share insights about the company's culture and compassion with the research participants for their review (Locke & Velamuri, 2009).

A shortcoming of this type of study is that researcher-selected photographs may not be meaningful for participants. To overcome this limitation, researchers should consider the perceived significance of their photos to participants before embarking on a study of this nature. Furthermore, this approach can be time-consuming and expensive. Although archival photos can generate discussion during interviews and focus groups, finding and obtaining permissions to use potentially copyrighted photographs from the popular press can be a lengthy and costly process. Despite these costs, using researcher-selected photographs can provide participants with a glimpse of how outsiders interpret organizational culture and can provide researchers with rich data from inside the firm about how participants interpret organizational items on public display.

Participant Photograph Study of Compassion as an Element of Organizational Culture

Proposed Study 4: To identify enduring cultural aspects of organizational compassion through participant-provided photographs, members could use cameras, either disposable or those already in their cellular phones, to take photographs of their culture. Participants would be given specific directions about what photographs to take and how to provide them to the researchers. Once the photographs are taken, researchers can choose which photos to discuss with organizational members either individually or in groups.

The researchers could identify similar photographs or patterns and then arrange these in group sessions for discussion. Having photographs from across different organizational units could aid comparisons. If different aspects of compassion are captured, these photographs are then grouped together for discussion. Once the photographs are grouped into categories, a group interview may result in deeper insights than interviewing alone because responding to photographs as a group can clarify thoughts and provide more depth and memory than individual interviews. A variant on this approach that an organization could undertake is what is referred to as a photovoice approach (Mitchell, 2011) to highlight the voice of marginalized organizational groups such as administrative staff, janitors, or part-time employees. In this approach, after the interviews are completed, the researcher selects photographs and quotes and provides feedback to those employees interviewed as possible management of the organization.

One consideration of this approach is the added time commitment of pilot testing. Researchers should pose their research question to a trial group of organization members to make sure that they understand the type of photographs to take. A possible shortcoming is that participants do not always take the task of photography seriously because it may seem like more of a fun activity than serious research. Even though the organizational members are asked to bring back meaningful photos, they may not produce usable data. This approach can also take a tremendous amount of time for the researchers, especially in selecting individuals for group elicitation and conducting interviews. Despite these shortcomings, allowing organizational members to create their own photographs that demonstrate organizational compassion and then discuss it with the researcher and other organizational members has the ability to lead to new theoretical insights about compassion and culture.

DISCUSSION

Photographic methodologies allow researchers to capture voices and stories from organizational members about a variety of topics including compassion and can thus provide value in several important ways. First, photographic methods allow the researchers to get close to lived experience of the organization and allow organizational members to tell their stories spontaneously (Collier & Collier, 1986). Photographs made by participants have the ability to "mine deeper shafts into a different part of human consciousness than do words-alone interviews" (Harper, 2002, p. 23). Second, organizational members may enjoy participating in this type of research, especially if they are given the task of taking pictures of their organization. Many participants have noted how enjoyable it is to take photographs and, according to several researchers, this involvement has improved their involvement and insights (Warren, 2002).

Risks and Caveats of Compassion Research

Whether photographs are archival, taken by a researcher, or taken by a participant, interviewing someone about compassion—either a specific event or in general in the culture—can be sensitive and intense. Interviewing with a photograph between a researcher and field participation can reduce the perceived power between the two people and allow the participant to relax, share more, and be more involved in collaboration. The main difficulty in implementing photographic methods in compassion research is that it may heighten the emotions beyond that of interviews alone or another aspect of an organization's culture. In fact, Collier and Collier (1986, p. 131) noted this in their research:

> Photographs are charged with unexpected emotional material that triggers intense feeling and divulges truth. It is probably more difficult to lie about a photograph than to lie in answer to a verbal question, for photographic scenes can cause intense feelings that are revealed by behavior, flushed faces, tense silence or verbal outbursts...The most innocent picture can create an explosion that changes the whole character of the interview.

Furthermore, we risk "insensitive intrusion with the camera," especially in studying compassion around particular events (Collier & Collier, 1986, p. 133). The emotional nature of photographic research methods (Collier & Collier, 1986), coupled with sensitive topics such as compassion, may also result in painful emotional labor (Hochschild, 1983) for researchers who are unprepared to deal with emotionally charged responses from participants. Asking participants to remember back to times of immense personal pain or to confront photographs that vividly remind them of sorrow places a tremendous burden on researchers as well as participants; thus, while the main concern of protecting human subjects centers around preventing harm to participants, researchers working on these types of topics with these types of methods may find themselves doubly at risk for "compassion stress" (Rager, 2005, p. 423).

Extending Photographic Methods Beyond Organizational Compassion

Against the backdrop of exploring organizational compassion, this level of participant involvement provides avenues for exploring other intangible concepts that permeate organizational life, including emotions, fairness, justice, and commitment. This method is suited to questions that seek to include a variety of voices in the conversation and it provides avenues for insights at various levels of analysis. Photographic methods can illuminate new findings in micro-level phenomena such as ethics, emotions, assumptions, decision-making, and individual practices (Chen, Treviño, & Humphrey, 2020). These inherently nebulous concepts can be made more concrete by using images of what they represent in facilitated interviews.

Additionally, photographic methods can drive discussions about meso-level phenomena such as leadership, loyalty, teamwork, and social structures. For example, research agendas that specifically look at differences between family- and nonfamily members within family businesses (Gottschalck, Guenther, & Kellermanns, 2019; Vincent Ponroy, Lê, & Pradies, 2019) may find photography a useful way to elicit responses from employees in both camps. Additionally, the concept of trust in and between organizations (Connelly, Crook, Combs, Ketchen, & Aguinis, 2018; Poppo, Zhou, & Li, 2016) can be more fully understood when participants are invited to create their own images of trust, either as it has been built or breached (Ferrin, Cooper, Dirks, & Kim, 2018).

Finally, photographic methods can provide a way for individuals to provide their input and insights on macro-level topics such as strategic change, industry dynamism, and firm rivalry. These concepts are typically the purview of executives and top managers (Nadkarni, Chen, & Chen, 2016; Wowak, Mannor, Arrfelt, & McNamara, 2016) however, there may be untapped insights from employees at other levels in the organizational hierarchy. By inviting participants to use photographs to illustrate concepts such as trust and dynamism, researchers and practitioners alike may discover untapped sources of empirical insight and theoretical inspiration. Table 2 provides an overview of other potential organizational phenomena that would benefit from photographic methods, their typical level of analysis, and potential research questions that participant- or researcher-generated photographs could begin to answer.

Table 2. Other Avenues for Photographic Research Methods.

Phenomenon	Level of Analysis	Research Questions
Loyalty and turnover	Micro	(1) What aspects of this organization demonstrate mutual loyalty between employees and the firm? (2) What unspoken employee issues are driving turnover in this firm or this field?
Emotions and cognitions	Micro	(1) What principles or elements guide ethical behavior? (2) What unchallenged assumptions drive decision-making?
Trust	Meso	(1) How is trust built or breached between organizational members? (2) How is trust built or breached between employees and the organization itself?
Family business issues	Meso	(1) What cues and signals do family- and nonfamily business members identify with? (2) How do nonfamily members see themselves in a family business?
Industry dynamism	Macro	(1) How has this business changed in the last [X] year time period? (2) What issues are likely to arise in the next [X] year time period?
Rivalry	Macro	(1) What elements of competition distinguish between successful and unsuccessful rivals in this industry? (2) What blind spots are executives overlooking?

CONCLUSION

At its core, compassion is an emotional and personal topic grounded in idiosyncratic needs and spontaneous responses (Frost et al., 2000). The inevitable yet unpredictable nature of human suffering means that pain is a constant aspect of life and organizational members constantly cope with a variety of personal tragedies. Photo elicitation has the ability to contribute to our understanding of how organizations care for their members during their darkest times and inform our theories of care, compassion, and culture.

REFERENCES

Banks, M. (2007). *Using visual data in qualitative research*. Los Angeles, CA: SAGE Publications.

Bateson, G., & Mead, M. (1942). *Balinese character*. New York, NY: New York Academy of Science.

Becker, H. S. (1974). Photography and sociology. *Studies in the Anthropology of Visual Communication*, *1*, 3–26.

Blackmar, F. W. (1897). The smokey pilgrims. *American Journal of Sociology*, *2*, 485–500.

Buchanan, D. A. (2001). The role of photography in organization research: A reengineering case illustration. *Journal of Management Inquiry*, *10*(2), 151–164.

Bushnell, C. J. (1902). Some social aspects of the Chicago stock yards. *American Journal of Sociology*, *7*, 289–330.

Cameron, K. S., Bright, D., & Caza, A. (2004). Exploring the relationships between organizational virtuousness and performance. *American Behavioral Scientist*, *47*, 766–790.

Chen, A., Treviño, L. K., & Humphrey, S. E. (2020). Ethical champions, emotions, framing, and team ethical decision making. *Journal of Applied Psychology*, *105*(3), 245–273.

Clark, C. (1997). *Misery and company: Sympathy in everyday life*. Chicago, IL: University of Chicago Press.

Collier, J., & Collier, M. (1986). *Visual anthropology: Photography as a research method*. Albuquerque, NM: University of New Mexico Press.

Connelly, B. L., Crook, T. R., Combs, J. G., Ketchen, D. J., Jr, & Aguinis, H. (2018). Competence- and integrity-based trust in interorganizational relationships: Which matters more? *Journal of Management*, *44*, 919–945.

Dacin, M. T., Munir, K., & Tracey, P. (2010). Formal dining at Cambridge Colleges: Linking ritual performance and institutional maintenance. *Academy of Management Journal*, *53*, 1393–1418.

Denzin, N., & Lincoln, Y. 2005. Introduction: The discipline and practice of qualitative research. In N. Denzin & Y. Lincoln (Eds.), *The SAGE handbook of qualitative research* (3rd ed.). Thousand Oaks, CA: SAGE Publications.

Dion, D. (2007). The contribution made by visual anthropology to the study of consumption behavior. *Recherche et Applications en Marketing*, *22*, 61–78.

Dutton, J., Frost, P., Worline, M., Lilius, J., & Kanov, J. (2002). Leading in times of trauma. *Harvard Business Review*, *80*, 54–61.

Dutton, J., Worline, M., Frost, P., & Lilius, J. (2006). Explaining compassion organizing. *Administrative Science Quarterly*, *51*, 59–96.

Edmondson, A. C., & McManus, S. E. (2007). Methodological fit in management field research. *Academy of Management Review*, *32*, 1155–1179.

Ferrin, D. L., Cooper, C. D., Dirks, K. T., & Kim, P. H. (2018). Heads will roll! Routes to effective trust repair in the aftermath of a CEO transgression. *Journal of Trust Research*, *8*, 7–30.

Frost, P., Dutton, J., Maitlis, S., Lilius, J., Kanov, J., & Worline, M. (2006). *Seeing organizations differently: Three lenses on compassion, Handbook of organization studies* (2nd ed.) (pp. 843–866). London: SAGE Publications.

Frost, P., Dutton, J., Worline, M., & Wilson, A. (2000). Narratives of compassion in organizations. In S. Fineman (Ed.), *Emotion in organizations* (pp. 25–45). London: SAGE Publications.

Gottschalck, N., Guenther, C., & Kellermanns, F. (2019). For whom are family-owned firms good employers? An exploratory study of the turnover intentions of blue-and white-collar workers in family-owned and non-family-owned firms. *Journal of Family Business Strategy* (in-press). doi: 10.1016/j.jfbs.2019.02.004

Guba, E. G., & Lincoln, Y. S. (1994). Competing paradigms in qualitative research. In N. K. Denzin & Y. S. Lincoln (Eds.), *Handbook of qualitative research* (Vol. 2, pp. 105–117). Thousand Oaks, CA: SAGE Publications.

Harper, D. (2002). Talking about pictures: A case for photo elicitation. *Visual Studies*, *17*, 13–26.

Harper, D. (2005). What's new visually? In N.K. Denzin & Y.S. Lincoln (Eds.), *Handbook of qualitative research* (pp. 747–762). Thousand Oaks, CA: SAGE Publications.

Hochschild, A. (1983). *The managed heart*. Berkeley, CA: University of California Press.

Kanov, J., Maitlis, S., Worline, M., Dutton, J., Frost, P., & Lilius, J. (2004). Compassion in organizational life. *American Behavioral Scientist*, *47*, 808–827.

Lee, T. W. (1999). *Using qualitative methods in organizational research*. Thousand Oaks, CA: SAGE Publications.

Lilius, J., Kanov, J., Dutton, J., Worline, M., & Maitlis, S. (2011a). Compassion revealed: What we know about compassion at work (and where we need to know more). In K. S. Cameron & G. Spreitzer (Eds.), *The handbook of positive organizational scholarship* (pp. 273–287). Oxford: Oxford University Press.

Lilius, J. M., Worline, M. C., Dutton, J. E., Kanov, J. M., Maitlis, S., & Frost, P. (2011b). Understanding compassion capability. *Human Relations*, *64*, 873–899.

Lilius, J., Worline, M., Maitlis, S., Kanov, J., Dutton, J., & Frost, P. (2008). The contours and consequences of compassion at work. *Journal of Organizational Behavior*, *29*, 193–218.

Lin, A. C. (1998). Bridging positivist and interpretivist approaches to qualitative methods. *Policy Studies Journal*, *26*, 162–180.

Locke, K., & Velamuri, S. R. (2009). The design of member review: Showing what to organization members and why. *Organizational Research Methods*, *12*, 488–509.

Madden, L. T., Duchon, D., Madden, T. M., & Plowman, D. A. (2012). Emergent organizational capacity for compassion. *Academy of Management Review*, *37*, 689–708.

Margolis, E., & Pauwels, L. (2011). *Visual research methods*. London: SAGE Publications.

Meyer, A. (1991). Visual data in organizational research. *Organization Science*, *2*, 218–236.

Mitchell, C. (2011). *Doing visual research*. London: Sage

Nadkarni, S., Chen, T., & Chen, J. (2016). The clock is ticking! Executive temporal depth, industry velocity, and competitive aggressiveness. *Strategic Management Journal*, *37*, 1132–1153.

Petersen, N. J., & Østergaard, S. (2004). *Organisational photography as research method: What, how and why*. New Orleans, LA: Research Methods Division, Paper presentation, Academy of Management.

Poppo, L., Zhou, K. Z., & Li, J. J. (2016). When can you trust "trust"? Calculative trust, relational trust, and supplier performance. *Strategic Management Journal*, *37*(4), 724–741.

Prosser, J. (1998). *Image-based research: A sourcebook for qualitative research*. Bristol, PA: Falmer.

Rager, K. B. (2005). Compassion stress and the qualitative researcher. *Qualitative Health Research*, *15*, 423–430.

Ray, J. L., & Smith, A. D. (2012). Using photographs to research organizations: Evidence, considerations, and application in a field study. *Organizational Research Methods*, *15*, 288–315.

Rynes, S. L., Bartunek, J. M., Dutton, J. E., & Margolis, J. D. (2012). Care and compassion through an organizational lens: Opening up new possibilities. *Academy of Management Review*, *37*, 503–523.

Shortt, H., & Warren, S. (2012). Fringe benefits: Valuing the visual in narratives of hairdressers' identities at work. *Visual Studies*, *27*, 18–34.

Stasz, C. 1979. The early history of visual sociology. In J. Wagner (Ed.). *Images of information: Still photography in the social sciences* (pp. 119–135). Beverly Hills, CA: SAGE Publications.

Sutton, R. I. (1997). The virtues of closet qualitative research. *Organization Science*, *8*, 97–106.

Van Maanen, J. (1979). Reclaiming qualitative methods for organizational research: A preface. *Administrative Science Quarterly*, *24*, 520–526.

Venkatraman, M., & Nelson, T. (2008). From servicescape to consumptionscape: A photo-elicitation study of Starbucks in the New China. *Journal of International Business Studies*, *39*, 1010–1026.

Vincent Ponroy, J., Lê, P., & Pradies, C. (2019). In a family way? A model of family firm identity maintenance by non-family members. *Organization Studies*, *40*, 859–886.

Vince, R., & Warren, S. (2012). Participatory visual methods. In C. Cassell & G. Symon (Eds.), *The practice of qualitative organizational research: Core methods and current challenges* (pp. 291–312). London: SAGE Publications.

Wagner, J. (1979). *Images of information: Still photography in the social sciences*. Beverly Hills, CA: SAGE Publications.

Warren, S. (2002). Show me how it feels to work here: Using photography to research organizational aesthetics. *Theory and Politics in Organizations*, *2*, 224–245.

Wowak, A. J., Mannor, M. J., Arrfelt, M., & McNamara, G. (2016). Earthquake or glacier? How CEO charisma manifests in firm strategy over time. *Strategic Management Journal*, *37*, 586–603.

BLOOMBERG SUPPLY CHAIN ANALYSIS: A DATA SOURCE FOR INVESTIGATING THE NATURE, SIZE, AND STRUCTURE OF INTERORGANIZATIONAL RELATIONSHIPS

Ace Beorchia and T. Russell Crook

ABSTRACT

Research involving interorganizational relationships (IORs) has grown at an impressive rate. Several datasets have been used to understand the nature and performance implications of these relationships. Given the importance of such relationships, we describe a relatively new dataset, Bloomberg SPLC, which contains data regarding the percentage of costs and revenues attributed to suppliers and customers, as well as allows researchers to construct a comprehensive dataset of IORs of buyer–supplier networks. Because of this, Bloomberg SPLC data can be used to uncover new and exciting theoretical and empirical implications. This chapter provides background information about this dataset, guidance on how it can be leveraged, and new theoretical terrain that can be charted to better understand IORs.

Keywords: Interorganizational relationships; alliances; supply chain; networks; buyer–supplier agreements; partnerships

INTRODUCTION

Over the last several decades, research involving interorganizational relationships (IORs) has grown at an impressive rate (cf. Connelly, Crook, Combs,

Advancing Methodological Thought and Practice
Research Methodology in Strategy and Management, Volume 12, 73–100

ISSN: 1479-8387/doi:10.1108/S1479-838720200000012017

Ketchen, & Aguinis, 2018; Parmigiani & Rivera-Santos, 2011). Research in this area has shown that firms operate in complex environments, and their fates are intertwined with those of other firms (e.g., Munyon, Jenkins, Crook, Edwards, & Harvey, 2019; Zimmermann & Foerstl, 2014). There are a variety of types of IORs, including strategic alliances (e.g., Ahuja, Lampert, & Tandon, 2008; Wassmer, 2010), joint ventures (e.g., Brouthers & Hennart, 2007), buyer–supplier agreements (e.g., Holcomb & Hitt, 2007; McCutcheon & Stuart, 2000), franchising (e.g., Combs & Ketchen, 2003), cross-sector partnerships (e.g., Selsky & Parker, 2005), networks (e.g., Hoang & Antoncic, 2003; Provan, Fish, & Sydow, 2007), interlocking directorates (e.g., Burt, 1980; Mizruchi, 1996), and coalitions (e.g.,Man Zhang & Greve, 2019; Polzer, Mannix, & Neale, 1998). Such relationships are generally considered "strategically important, cooperative relationships between a focal organization and one or more other organizations to share or exchange resources with the goal of improved performance" (Parmigiani & Rivera-Santos, 2011, p. 1109).

While research supports the notion that firms' fates are linked with other firms (Parmigiani & Rivera-Santos, 2011), IOR data can be difficult to procure. There are several reasons for this. First, firms may prefer to keep such relationships private in order to maintain a competitive advantage (Handfield, 2012). Second, few regulations require that firms disclose their IORs. In fact, only public firms are required – at least in the United States – to disclose a relationship when a customer is responsible for at least 10% of a firm's revenue on an annual basis (Davenport, 2011). Of course, this list of customers is relatively small for most firms – containing no, one, or a just few key relationships. Due of the lack of available data on IOR linkages, relatively few data sources are available that allow researchers to study such relationships among firms, such as SDC Platinum (e.g., Kim, Howard, Cox Pahnke, & Boeker, 2016), ReCap (e.g., Mindruta, Moeen, & Agarwal, 2016), and ISS (formerly RiskMetrics) (Sauerwald, Lin, & Peng, 2016).

Although these datasets can provide valuable insights into IORs – mainly strategic alliances, networks, and board interlocks – data sources quantifying the size of IORs have been largely unavailable. Considering the magnitude of IORs has important theoretical implications. It is well known that relationship type (e.g., alliance, network, joint venture) is an important factor of IORs (Barringer & Harrison, 2000). However, it is also known that each relationship and type that firms engage in can have varying levels of magnitude (Golicic, Foggin, & Mentzer, 2003) that impact how and to what extent a firm interacts with other firms. For example, although a firm may have multiple suppliers and customers, it is likely to allocate more resources to suppliers and customers to which the central firm is more exposed (Golicic et al., 2003). Up to now, researchers have had limited access to data regarding IOR magnitude – especially from secondary data sources.

In this chapter, we describe a relatively new data source that has been largely unexplored by management scholars (Kim & Davis, 2016) – the Bloomberg

Supply Chain Analysis (hereafter Bloomberg SPLC) – and how it may be used by researchers interested in IORs.[1] With data dating back to 2006, the SPLC function in Bloomberg allows users to search a focal firm and view a list of its suppliers, customers, and competitors. Specifically, it offers data regarding the percentage of costs and revenues attributed to suppliers and customers. This allows researchers to create a comprehensive dataset of IORs to construct entire buyer–supplier networks. When paired with other datasets such as Compustat and Factiva, among others, Bloomberg SPLC data have the potential to yield new insights into the nature of IORs. In particular, we believe the use of Bloomberg SPLC data will allow researchers to more effectively theorize about the implications of the magnitude of IORs and use Bloomberg SPLC data to test and expand theoretical perspectives, mainly transaction cost economics (TCE), resource-based (relational) view, resource dependence theory (RDT), stakeholder theory, institutional theory, and social network theory.

BACKGROUND ON IOR RESEARCH

All firms cooperate with other organizations for their survival and growth (Scott, 1987). However, the relationships entered into vary both in *why* and *how* firms decide to cooperate (Oliver, 1990; Parmigiani & Rivera-Santos, 2011). Put differently, the intent as well as the structure of cooperation has historically been important to IOR researchers. In terms of motives, researchers suggest that firms enter into IORs in an effort to explore or exploit knowledge, tasks, functions, or activities (March, 1991). Parmigiani and Rivera-Santos (2011) refer to these motivations as "co-exploration" and "co-exploitation."

Under the assumption that firms enter these relationships intentionally and consciously for a strategic purpose, several determinants of IORs also help to explain the motivation for cooperation. According to Oliver (1990), these motivations include necessity (i.e., to meet legal or regulatory requirements – Zald, 1978), asymmetry (i.e., to assert power, influence, or control over organizations with scarce resources – Crook & Combs, 2007), reciprocity (i.e., to pursue common goals or interests – Jamali & Keshishian, 2009), efficiency (i.e., to improve internal input/output ratio – Elston, MacCarthaigh, & Verhoest, 2018), stability (i.e., to reduce environmental uncertainty – Park & Mezias, 2005), legitimacy (i.e., to appear in agreement with prevailing norms, rules, beliefs, or expectations – Dacin, Oliver, & Roy, 2007), or some combination of these motivations.

Beyond these motivations, IORs can also yield benefits by helping firms gain access to resources, build economies of scale, share risks and costs, access new markets, develop product and services, learn, reduce time to market, become more flexible, lobby as a collective, and block competitors, among others (Barringer & Harrison, 2000). Taken together, the list of motivations and benefits

[1]Recently several researchers have published articles in top management and supply chain management journals utilizing Bloomberg SPLC (i.e., Elking et al., 2017; Kim & Davis, 2016; Steven et al., 2014).

of engaging in IORs is extensive. So, too, are the types of IORs. As noted earlier, firms engage in various IOR types, including strategic alliances, joint ventures, buyer–supplier agreements, franchising, cross-sector partnerships, networks, trade associations, interlocking directorates, and coalitions (e.g., Barringer & Harrison, 2000; Oliver, 1990).

Researchers interested in IORs have applied a variety of theoretical perspectives to understand motivations and benefits. From an economic standpoint, researchers have theorized that firms engage and cooperate with other organizations to become more efficient, reduce costs, acquire resources, and gain economies of scale (cf. Barringer & Harrison, 2000). In doing so, researchers have leveraged ideas from TCE (i.e., IORs used to increase efficiency), and resource-based view (RBV) (i.e., IORs used to acquire new resources), to name a few. From an organization theory standpoint, researchers have noted the importance of social structures and relationships in IORs such that firms engage in them to increase social standing, make connections, flex power, as well as reduce dependency and uncertainty. In doing so, researchers have built on ideas from RDT (i.e., to reduce uncertainty due to power and dependence), stakeholder theory (i.e., to reduce uncertainty from reputational concerns), institutional theory (i.e., to conform to socially constructed norms), and social network theory (i.e., to gain information and knowledge by creating or strengthening a relationship) (cf. Parmigiani & Rivera-Santos, 2011).

The topic of IORs remains a vibrant area of inquiry within strategic management and organizational theory research (e.g., Oliveira & Lumineau, 2019; Villena, Choi, & Revilla, 2019). Indeed, to obtain a relevant understanding of the most current IOR research and datasets being examined, we reviewed articles published from 2016 to 2018 in *Strategic Management Journal*. We flagged articles researching IORs, and identified the dataset(s) used in each study (Table 1). In this review, we discovered that all IOR types are not given equal attention in current research, an occurrence others have also noted (e.g., Parmigiani & Rivera-Santos, 2011). Specifically, recent studies have largely focused on interfirm alliances (e.g., Asgari, Singh, & Mitchell, 2017; Bakker, 2016; Blevins & Ragozzino, 2018) and networks (e.g., Jiang, Xia, Cannella, & Xiao, 2018; Kim et al., 2016; Sauerwald et al., 2016). Moreover, many studies rely on the same data sources in IOR research. For example, SDC Platinum and ReCap – as described earlier – are two of the most commonly used data sources in recent strategy research.

IOR studies using SDC Platinum suggest that predictors of alliance formation are determined by the context in which firms are situated (i.e., industry, time period) (Ghosh, Ranganathan, & Rosenkopf, 2016), that the presence of venture capitalists increase entrepreneurial firms' alliance formation (Blevins & Ragozzino, 2018), and that firms in cohesive alliance networks tend to search outside the network for new partners (Jiang, Xia, Cannella, & Xiao, 2018). Additionally, studies using ReCap have found that some firms are not able to ally with their most preferred partners (Mindruta et al., 2016), that firms reconfigure their alliance portfolio following a technological discontinuity (Asgari et al., 2017), and that administrative controls can regulate knowledge transfers across research and development alliances (Devarakonda & Reuer, 2018).

Table 1. Recently Used Interorganizational Relationship (IOR) Secondary Datasets in the *Strategic Management Journal.*

Dataset	IOR Types	Citations	Complementary Datasets Utilized	Content
SDC Platinum	Alliances, networks	Blevins and Ragozzino (2018), Bos, Faems, and Noseleit (2017), Devarakonda and Reuer (2018), Ghosh et al. (2016), Howard, Withers, Carnes, and Hillman (2016), Jiang et al. (2018), and Kim et al. (2016)	BioScan Directory, Bloomberg Private Firm Directory, Bureau van Dijk, Center for Research on Securities Pricing (CRSP), Compustat, Execucomp Fortune's Most Admired Firms, Institutional Brokers' Estimate System (IBES), News archives, Patent Network Dataverse, Proxy statements, United States Patent and Trademark Office (USPTO), ReCap, U.S. Securities and Exchange Commission (SEC)	• Alliance information • Mergers and acquisition data • Joint venture and repurchase information • Corporate governance data (e.g., shareholder rights plan adoptions, amendments, and expirations; dissident shareholder campaign data)
ReCap	Alliances, networks	Asgari et al. (2017), Jiang et al. (2018), Devarakonda and Reuer (2018), Kim et al. (2016), and Mindruta et al. (2016)	BioScan Directory, Factiva, SDC Platinum, European Patent Office (EPO), PATSTAT	• Pharmaceutical (biotech) alliance (i.e., R&D) data
Compustat	Alliances, networks	Howard et al. (2016), Kim et al. (2016), and Mindruta et al. (2016)	Center for Research on Securities Pricing (CRSP), SDC Platinum	• Firm size and financial measures (e.g., market return data, assets, ROA)
Directory of Corporate Affiliations (LexisNexis)	Alliances	Mindruta et al. (2016)		• Firm information (e.g., description, financials, classifications) • Competitor names • Executive and board of director data • External service firms • Company hierarchy information
Factiva	Alliance	Asgari et al. (2017)	ReCap	• Qualitative data about firms (e.g., press releases, company reports, newspaper articles,

Table 1. *(Continued)*

Dataset	IOR Types	Citations	Complementary Datasets Utilized	Content
				journal publications, newswires, TV or radio podcasts)
BioScan Directory	Alliances, networks	Jiang et al. (2018) and Mindruta et al. (2016)	ReCap, SDC Platinum	• Biotechnology alliance data
Register of Australian Mining	Alliances	Bakker (2016)		• Data on Australian mining companies, directors, and exploration projects
RiskMetrics Directors Universe (Now ISS)	Networks	Sauerwald et al. (2016)		• Detailed board of director information (e.g., name, age, tenure, gender, committee memberships, primary employer and title) • Voting results and analytics • Governance data (e.g., classified boards, cumulative voting) • Executive compensation data
US Patent and Trademark Office (USPTO)	Buyer–supplier relationships	Mawdsley and Somaya (2018)	SDC Platinum	• Patent information (e.g., name, information, year, type, dates, processing time) • Inventor, coinventor, and assignee information (e.g., name, geographic location, date of most recent patent)

Although these data sources have helped researchers yield a variety of important insights, our review also showed that the most common and available data sources contain information on strategic alliances and networks, which in turn, represent the most commonly studied IORs. As a result, most IOR types and datasets in our review (see Mawdsley & Somaya, 2018 for an exception) focus on horizontal rather than vertical relationships. Yet, because vertical relationships are what comprise modern day supply chains (Shook, Adams, Ketchen, & Craighead, 2009), we suggest that IOR research – at least involving vertical relationships – would benefit from datasets that provide information about such relationships among organizations.

NEW DATA SOURCE FOR IOR RESEARCH

We introduce a relatively new and virtually untapped data source, Bloomberg Supply Chain Analysis, that can facilitate data collection regarding both horizontal and vertical ties, show the size of these ties, and, ultimately, open the door for new insights into IORs. According to Bloomberg, SPLC

> ...provides a comprehensive supply chain breakdown for a selected company, so one can analyze revenue exposure for the focal firm, its suppliers, and its customers, as well as track the performance of a company against its peers. (Bloomberg Professional Services, 2017)

Tables 2 and 3 highlight key information available in the dataset. Specifically, Bloomberg SPLC reports a firm's key relationships with suppliers and customers, as well as provides information on geographic areas and industries. In addition, Bloomberg SPLC reports the size of relationships among the firms by listing the actual or estimated percent of cost and revenue of suppliers and customers; this information is valuable as it not only provides whether ties exist, but also the size and strength of the ties – data often missing in commonly used datasets. Lastly, Bloomberg SPLC can enrich our understanding of IORs because it contains buyer–supplier relationships dating from 2006 to the present; this provides an opportunity for researchers to analyze how IOR dynamics change over time for one firm, dyads, or complete buyer–supplier networks over 10 years, in some cases.[2]

Although Bloomberg SPLC data alone appear capable of providing new insights into IORs, we also see opportunities to leverage these data together with

Table 2. Bloomberg SPLC Data for Apple (APPL) Quantified Suppliers as of 12/31/2018.

ID	Name	Country	Supplier/ Customer	% Revenue	% Cost	Relationship Value (USD)	Value Units	Account As Type	Source	As of Date
1	Hon Hai Precision Industry Co., Ltd	TW	Supplier	45.23	52.28	20.29	Billion	COGS	Estimate	12/12/2018
2	Quanta Computer Inc.	TW	Supplier	62.01	14.70	5.71	Billion	COGS	Estimate	11/16/2018
3	Pegatron Corp.	TW	Supplier	64.44	14	5.60	Billion	COGS	Estimate	11/06/2018
4	Foxconn Industrial Internet Co., Ltd	CN	Supplier	27.17	10.29	3.27	Billion	COGS	Estimate	09/04/2018
5	Samsung Electronics Co., Ltd	KR	Supplier	5.97	8.97	3.48	Billion	COGS	2018 A CF	12/14/2018

Source: Bloomberg Finance L.P. Used with permission of Bloomberg Finance L.P.

[2]Again, we highlight at while data are available back to 2006, they are often less comprehensive prior to 2012.

Table 3. Variables Available in Bloomberg SPLC.

Variable	Description
Name	Name or ticker of the suppliers or customers. The focal firm of focus is always listed first in the table.
Country	Country of the company being analyzed.
% Revenue	• Supplier relationships: Percent of revenue a supplier receives from the focal firm (i.e., what percent of the supplier's revenues does the focal contribute). • Customer relationships: Percent of revenue the focal firm receives from the customer (i.e., what percent of the focal firm's revenues does the customer contribute).
Relationship Value	• The monetized value of the dyadic relationship (i.e., what amount of money is exchanged in the relationship).
Account As Type	The cost category to which Bloomberg assigns the transactions of a relationship (i.e., COGS, CAPEX, SG&A, R&D).
% Cost	• Supplier relationships: The percentage of the focal firm's costs that goes to the supplier (i.e., what percentage of the focal firm's costs are attributed to the supplier). • Customer relationships: The percentage of a customer's costs that go to the focal firm (i.e., what percentage of the customer's costs are attributed to the focal firm).
Source	The accounting period of the source, if the data was disclosed. When the data is estimated, this is indicated in this column. Users can click this variable to open the source documentation, when available.
As Of Date	Date that represents the publishing date of a company report containing the relationship data, or the date when the relationship estimate was updated.

information from other datasets. In doing so, this "combined" dataset will allow less common IOR types to be tested. For example, by combining the buyer–supplier data from Bloomberg SPLC with political action committee data (i.e., political donations by firms and employees within firms), researchers can better understand how supply chains develop coalitions to shape institutions that have direct and indirect influences on firms' abilities to survive and grow. In this case, the dataset would allow researchers to investigate independent and combined effects of different IOR types (i.e., buyer–supplier relationships and coalitions) to provide a more nuanced assessment of how IORs work alone and together in today's business environment. Other examples might include pairing the Bloomberg SPLC data with Compustat data to investigate board interlocks in vertical networks or with qualitative data from company websites, presentations, or statements to study when an organization seeks to boost or preserve its reputation, status, or celebrity by highlighting important vertical relationships in a network. Overall, the data appear extremely useful in testing theory and examining boundary conditions for a wide range of IORs. The next section describes the contents of the Bloomberg SPLC data source and instructions how to extract the data to create a custom dataset.

Bloomberg SPLC

Bloomberg SPLC is a comprehensive dataset of buyer–supplier data from 2006 to the present – although the dataset is more complete starting around 2012 (i.e., a firm may only show four suppliers in 2006 but dozens in 2012). As of 2011, Bloomberg SPLC covered buyer–supplier relationships for over 35,000 firms and source documents in over 12 languages (Davenport, 2011; Kim & Davis, 2016). The database facilitates the mapping of a firm to its customers, suppliers, and competitors, and it contains variables describing the supply chain exposure of the firms (Table 2). Additionally, revenue exposure for the focal firm, suppliers, and buyers is provided in the dataset so that money flows can be traced through a network. Broadly speaking, the dataset can help answer the following questions (Davenport, 2011):

- What firms are vertically connected in buyer–supplier relationships?
- Who are a firm's largest customers and suppliers? To what firms is a focal firm most exposed to?
- What are the sizes of these relationships, and do they present risks?
- How have a firm's customers and suppliers performed along recent financial results?

Because of few mandated regulations requiring public companies to disclose buyer–supplier relationships, vertical IOR data for firms is difficult to procure. While the SEC Regulation S-K "requires disclosure of the name of any customer that represents 10 percent or more of the issuer's revenues and whose loss would have a material adverse effect on the issuer" (Aleman, 2018, p. 87), there are no other supplier or buyer disclosure requirements, and other countries lack even this level of relationship disclosure (Davenport, 2011). Therefore, if researchers rely solely on the publicly disclosed relationships of firms, our understanding of IORs would be incomplete – missing a majority of vertical relationships.

To compensate for this lack of data, Bloomberg SPLC relies on a human and computerized process to aggregate publicly disclosed data. Moreover, Bloomberg SPLC has developed advanced proprietary supply chain data system to search and calculate, either mathematically or algorithmically, more complete buyer–supplier relationship information (Steven, Dong, & Corsi, 2014). In other words, Bloomberg SPLC will discover buyer–supplier relationships, record any disclosed relationship exposure data, and then take any unquantified relationships (i.e., relationships with an unknown numerical exposure) and convert them to quantified relationships (i.e., relationships with a known numerical exposure, such as sales). To do this, first Bloomberg identifies disclosed supply chain data from various sources, such as public filings (i.e., annual or other periodic reports), conference call transcripts, capital markets presentations, sell-side conferences, company press releases, and company websites (Davenport, 2011). A Bloomberg employee then determines for each relationship whether the product or service being sold is accounted by the customer as a cost of goods sold (COGS), capital expenditure (CAPEX), research and development (R&D), or selling, general, and administrative expenses (SG&A).

Next, if a figure quantifying a buyer–supplier relationship is publicly available, Bloomberg SPLC calculates the buyer–supplier exposure. For example, consider the following example provided by Bloomberg:

> When INTC [Intel Corporation] tells us in its 10-K that it receives 21% of its revenues from HPQ [HP INC], we identify the fact that INTC is selling semiconductors to HPQ, and that HPQ accounts for these semiconductors as a COGS item on its income statement. We then multiply 21% times INTC's revenues and divide that by HPQ's COGS, and derive that INTC represents 9.5% of HPQ's COGS. This number, while proprietarily derived and currently unique to Bloomberg, is nevertheless factual, insofar as the number that INTC gives us (21%) is correct. (Davenport, 2011, p. 2)

For these relationship exposures calculated mathematically, Bloomberg provides the source (e.g., annual report), the date of the source (e.g., 2018 Q3), and a link to view the source.

In situations where quantified relationship exposure data are not available and thus cannot be calculated, Bloomberg relies on its Bloomberg's Supply Chain Algorithm (BSCA) to algorithmically derive an estimate. Through this method, Bloomberg SPLC first takes into account numerous data sources, including quantified and unquantified relationships, accounting types, financials, geographies, end markets, operating segments, products, channels, and a variety of industry data (Kim & Davis, 2016). Next, it limits the range of possible values by creating a ratio between the supplier's revenue and the customer's costs to create a mathematically impossible area. For example, consider that company A is a supplier of company B with an unknown exposure relationship. If company A's total revenue is $100,000, and company B's total costs are $10,000,000, BSCA calculates that company A cannot compose more than 10% of company B's costs.

BSCA then continues taking other data into account to further narrow down the potential relationship. For example, other revenues or COGS unrelated to this relationship, known quantified relationships reported in public filings, other known distributors, and additional sources that help estimate the final relationship magnitude are considered in the algorithm in an effort to come up with an accurate estimate of the relationship. The following is an example of a relationship derived using BSCA provided by Bloomberg:

> Take, for example, AAPL [Apple] and T [AT&T]. Many people know that T is a customer of AAPL, but AAPL nor T disclose the size of the relationship. Bloomberg's algorithm determines that AAPL receives 5.11% of its revenues from T, and further that AAPL accounts for 9.49% of T's COGS. Further, we plan to provide statistical confidence intervals to provide users the range for which we have high confidence in which this single-point estimate resides. (Davenport, 2011, p. 2)

Bloomberg SPLC is constantly updating supply chain data on a daily basis based on the most current information they have available. If a company has gone one year without disclosing updated relationship information, Bloomberg

SPLC offers the company a window of a few weeks to update the information. If the company does not update the information, the relationship "expires" and changes to an unquantified status. For companies that disclose quarterly relationships, the system will update the information quarterly. The algorithms are also updated on a quarterly basis.

Bloomberg SPLC Contents and Extracting Information

Bloomberg SPLC data are available in two views that offer different benefits and drawbacks: chart view and table view. After entering the SPLC function in Bloomberg and inputting a focal firm to analyze, the chart view will display as the default (Appendix 1). In this view, a diagram is depicted with the focal firm in the middle of the screen, a list of suppliers on the left-hand side of the screen, and a list of buyers on the right-hand side of the screen. By default, the suppliers and customers are sorted by "company exposure." The suppliers are sorted by the cost paid by the focal firm, and the customers are sorted by the revenue the focal firm gets from the customers (i.e., to which suppliers and buyers is the focal firm most exposed). However, the suppliers and customers may also be sorted by "relationship exposure," which will list suppliers in descending order to revenues received from the focal firm and customers in descending order as a percentage of the focal firm's costs from their perspective (i.e., which suppliers and buyers are most exposed to the focal company). Finally, a custom sort using other criteria is also available. In the chart view, below the focal firm, a list of key competitors is also provided.

Listed below each supplier in the chart view is exposure data related to the focal firm. Specifically, the revenue percentage represents the percentage of the supplier's revenue the supplier receives from the focal firm. The cost percentage represents the percentage of the focal firm's costs that is accounted for by the costs incurred from the suppliers. By clicking on an individual supplier, a window will appear that contains more detail of the relationship, including relation exposure represented as a dollar amount, the accounting type used (i.e., COGS, CAPEX, R&D, SG&A), whether the exposure data are from a source or an estimate, and if from a source, the link to the document. In short, these percentages describe the magnitude of buyer–supplier relationships and also offer an objective assessment of exposure and risk.

Customers of the focal firm are listed on the right-hand side of the screen. Similar to the suppliers, the customers also have exposure data listed that relate to the focal firm. However, for the supplier, the revenue percentage represents the percentage of the focal firm's revenue the focal firm receives from the customer. The cost percentage represents the percentage of the customer's costs that is accounted for by the costs incurred from the focal firm. Again, by clicking on a customer, a window will appear containing similar information as the supplier.

Additionally, in the chart view, above the network diagram is the control area. In this space, researchers have the ability to select a currency, a display (i.e., company name or ticker), and whether the researcher desires quantified (i.e., relationships with an actual or estimated exposure data), unquantified (i.e., relationships without actual or estimated exposure data), or government contracts (i.e., subcontractors or government agencies and primes) displayed. Also, in this area, researchers can filter the data based on various variables (e.g., country of domicile, sector). Researchers can also select an "as of date" to determine the date to which the data correspond. For example, if 12/31/2018 is selected, the data will represent the most accurate and updated relationship data as of that day. Finally, using the "analyze" icon (only available for same-day data), researchers may select several financial figures from Bloomberg with which to analyze the supply chain by price change, revenue expectations, inventory growth, and more (e.g., price change %, latest sales surprise %, latest guidance to estimate difference %, projected sales growth, latest inventory growth).

While in the chart view, to extract the data, researchers would need to hand type the data into a new database software, such as Excel. If more detailed information other than the visible buyer/supplier and exposure percentages is desired (e.g., dollar amount of the exposure, whether the exposure is estimated or from a source), each customer or supplier would need to be individually clicked and data from the pop-up window copied. Thus, while the chart view is aesthetically pleasing, and visualizing the buyer–supplier relationships in this format may be convenient, the table view is more conducive to gathering all of the buyer–supplier data Bloomberg SPLC offers more efficiently.

Clicking on the "table" button in the control area of the screen reveals the same data from the chart view, only presented in a table with all information neatly displayed (see Appendix 2). In this view, the control area presents similar options as in the chart view (e.g., select a currency, display names or tickers, display quantified or unquantified relationship, create custom filters). However, because only customers or suppliers are listed in the table, one must select the relationship type to display in this area (i.e., customer, supplier, peer, government) from the control area at the top of the screen. Important buyer–supplier variables are neatly listed in the table view, and researchers can choose additional financial data such as valuation metrics, liquidity measures, or other data available on the Bloomberg terminal to add to the table. Again, in the control area, researchers can select the "as of date" to choose the date with which the data should correspond. This feature can be useful when collecting longitudinal data or ensuring all data are pulled as of the same date.

Below the control area is the table itself. In the table, the top row is the focal firm being analyzed, and the other rows represent customers or suppliers, depending on the relationship selected. Like the chart view, the customers or suppliers are listed by company exposure as default, but by adjusting settings in the control panel or right clicking on table headings, the display order can be adjusted. The columns in the table include the company name or ticker, the country of the company, the percent of revenue, the relationship value, the account type, the percent of cost, the source of the data (i.e., company disclosure or estimate), and the source's date. Using the table view, researchers can more easily copy necessary information into a secondary software such as Microsoft

Excel to create a dataset. As of this writing, Bloomberg does not allow SPLC data to be copied from their software, thus the information from the table will likely need to be hand-copied. However, according to their help documents, Bloomberg offers excel integration for individuals with a Bloomberg Anywhere license which allows the top 20 suppliers and customers to be exported from their program (Help Page, 2018).

When creating a dataset using Bloomberg SPLC, researchers may want to include an additional column to mark each unique relationship beginning with one and ending with the last relationship. Additionally, if creating a dataset of both customers/buyers and suppliers, it is important to create a variable in the dataset that codes the relationship accordingly. For example, if we were creating a longitudinal dataset of Apple's supply chain, we would first enter the SPLC function and type APPL as the focal firm we are interested in. We would then navigate to the "as of date" and select 12/31/2018. We would select "Quantified Suppliers" as the view and switch the display to tickers. Not only does using a ticker increase efficiency in copying the data but it is also beneficial to use the ticker if the data are to be paired with other datasets. We would then copy the data (in this case, the top 10 suppliers) from Bloomberg SPLC into a Microsoft Excel spreadsheet. We would add an additional column to give each relationship a unique identifying number and a column to code the relationship as a supplier relationship. Again, this is vital since the variables represent different information depending on the relationship type. Next, we would enter 12/31/2017 as the new "as of date" and copy over the updated information. If we were pulling 10 years of data, we would continue until 12/31/2009, for example.

Next, we would select the "Quantified Customers" as the view and follow the same steps above but for customers rather than suppliers. Once completed, our dataset would include all of Apple's top 10 suppliers and customers from the past 10 years with the associated exposure data for each relationship. As previously mentioned, without a Bloomberg Anywhere license, the software prohibits copying of the data to a clipboard (i.e., no copying and pasting data), and the data must be copied by hand. However, users with a Bloomberg Anywhere license may take advantage of the Excel Integration feature of SPLC and can easily export the top 20 suppliers by spend and revenue as well as customers by revenue (Help Page, 2018).

Pairing with Other Data Sources

Using the firms' ticker symbols common identifiers, the Bloomberg SPLC data's value increases dramatically when paired with other datasets. This can be done through several processes. First, a sample of organizations can be identified using an existing data source, and buyer–supplier relationship data could be collected from Bloomberg SPLC for the specific firms in the sample. For example, utilizing a list of all public companies who have donated to a political action committee (PAC), buyer–supplier relationship data can be paired with the PAC data to test hypotheses regarding IORs such as coalitions or strategic alliances. Next, a sample of firms could be selected based on specific characteristics (e.g., industry, geography, size) and Bloomberg SPLC data could be compiled for firms within this sample. Next, additional data from secondary sources could be used to test hypotheses specific to those groups. For example, industry networks could be compiled through Bloomberg SPLC and paired with board of director data to test hypotheses relating to board interlinks in vertical IORs. And finally, Bloomberg SPLC data can be paired with qualitative data (e.g., public reports, website information, mission statements) to test questions about firms' discourses about suppliers and buyers, and to what extent firms exhibit homophily in the supply chain in terms of mission statements and CSR initiatives. While these are only a few examples, we believe researchers will find creative ways to utilize Bloomberg SPLC in conjunction with other data sources.

POTENTIAL INSIGHTS FROM SPLC DATA

In this section, we first share a study that may be complemented through the use of Bloomberg SPLC data. Next, we share several areas in which the data provided by Bloomberg SPLC can add to the research base in management. Relying on an established list of theoretical motivations for the existence of IORs (Parmigiani & Rivera-Santos, 2011), we demonstrate how Bloomberg SPLC can be a useful resource to expand theory and answer important research questions (Table 4). By briefly reviewing these theories and suggesting new ways that Bloomberg SPLC can be used to study IORs, we hope to inspire researchers to use this dataset as a tool to augment research of IORs from various perspectives.

Application of Data

Recently, Kim and Choi (2018) examined the strategic value of strong ties, weak ties, and intermediate ties. They theorize that, unlike strong ties that can provide strategic value through well-established routines and information sharing, and weak ties that can provide access to innovative information and/or valuable resources, the intermediate ties lack any strategic benefit and may even increase cognitive burdens and coordination costs (Kim & Choi, 2018). Hence, the authors hypothesize that there is a U-shaped relationship between the strength of buyer–supplier ties and value creation. Next, the authors also hypothesized that dependence asymmetry, or an imbalance of interfirm dependencies, between buyer and supplier partnerships would result in decreased joint value creation. In addition, they believed that this negative impact of dependence asymmetry would have a more pronounced negative effect on intermediate ties.

To collect data, the authors created a survey for a North American automaker and its suppliers. This survey was administered to the firm's suppliers and was cross-validated through a sample of buyer-side surveys. The survey contained measures of value creation, strength of the tie, and dependence asymmetry. The

Table 4. Theoretical Questions Potentially Answered Using Bloomberg SPLC Data.

Theory	Example of Research Questions
Transaction cost economics	Do firms with fewer suppliers with higher exposures outperform firms with many suppliers with fewer exposures? Under what conditions and with what outcomes do firms use plural sourcing partners? How do firms create efficiencies in their supply chains? How are efficiencies created over time through vertical IORs?
Resource-based (relational) view	Are firms that transact with one another at greater levels able to build more relational resources? Do resources obtained from vertical IORs accumulate more on the buyer or supplier side? What environmental landscape shapes the nature of resource accumulation for vertical IORs? Do firms that bridge structural holes in buyer–supplier networks have a competitive advantage that results in increased performance? Do firms who participate in tight networks of interlinked buyers and suppliers outperform competitors? What type of network characteristics provide firms with competitive advantages? How do these competitive advantages change over time?
Resource dependence theory	How do firms minimize power and dependence gaps with their buyers and/or suppliers? How and under what conditions do vertical IORs combine resources to maximize power over or minimize dependence on other organizations or institutions?
Stakeholder theory	Do firms with powerful customers and suppliers explicitly communicate these relationships to other shareholders in an attempt to increase legitimacy? How do suppliers and/or buyers respond to a firm whose legitimacy has decreased? How do relationship exposures change for buyers and suppliers after a negative event? Do value-driven firms create vertical IORs with firms with similar values?
Institutional theory	Do corporate social responsibility initiatives of large firms in a supply chain trickle up or down the vertical relationships? Do smaller firms engage in normative or mimetic actions to adhere to the behavior of more legitimate firms in their buyer–supplier networks? When does a firm decouple or defy traditional rational explanations within a network or industry?
Social network theory	How do buyer–supplier networks change over time? Does network location influence firm performance within buyer–supplier networks? How does a negative event between two firms impact other relationships in the social network? Under what conditions and with what impact do board members interlock in buyer–supplier networks? How do buyer–supplier networks differ by industry or geographic qualities?

results widely supported the authors' hypotheses and revealed the importance of considering both tie strength, dependence, and intermediate ties in future research.

We believe this study adds an important piece to our knowledge of buyer–supplier relationships. Yet, at the same time, using the Bloomberg SPLC dataset could provide additional value to the researchers and build upon the limitations identified by the authors. Bloomberg SPLC is a particularly relevant data source to test containing measures of tie magnitude and strength (i.e., percentage of cost and revenue exposure) of the buyer and supplier as well as dependence asymmetry (i.e., incongruity between the cost and revenue exposure for buyer and supplier). When paired with other data on firm performance, the Bloomberg SPLC aligns nicely with Kim and Choi's (2018) model.

Kim and Choi mention several limitations to their study. First, they state that response bias may influence their findings. Because Bloomberg SPLC contains objective data, response bias would not influence some key facets of their findings if using the dataset. Next, while their method allowed the researchers to conduct rich and context-specific data, they state that their results may not be generalizable outside of their particular context since they only surveyed the suppliers of one firm. Bloomberg SPLC contains buyer–supplier information for tens of thousands of firms and could provide more generalizable results. Finally, they propose that absorptive capacity may be another possible moderator in their study, but they lacked information on the secondary relationship levels (i.e., the suppliers of the supplier and the buyers of the buyer) in the supply chain to test this relationship. Of course, this information could also be gathered using Bloomberg SPLC.

Overall, Kim and Choi's study adds to our understanding of supplier–buyer relationships through extrapolating on dependence asymmetry and intermediate ties. By conducting a similar study utilizing Bloomberg SPLC data, we believe the findings may help resolve some of the stated limitations of the study by removing survey bias, allowing for the testing of other moderators, and providing a sample that would allow for more generalizability. In addition, we believe Bloomberg SPLC could be used to further break apart this notion of intermediate ties. For example, using longitudinal data from the dataset, researchers could investigate under what conditions intermediate ties become strong or weak. Additionally, the relationship between the strength of ties and the dependence symmetry or asymmetry of buyer–supplier partnerships could also be further explored using the Bloomberg SPLC data.

Transaction Cost Economics

From a TCE perspective, IORs and firms are viewed as alternative governance structures that can be used to organize transactions (Williamson, 1985). TCE claims that the characteristics of transactions – in part – shape the extent to which governance structures can be used to efficiently organize transactions. Efficiency, in TCE terms, is a function of the transaction costs that arise; such expenses include identifying market prices, negotiating, and carrying out the exchange

(Williamson, 1985). And, there is broad evidence that supports TCE's assertion that governance efficiency is an important determinant of how transactions are organized (cf. Crook, Combs, Ketchen, & Aguinis, 2013). Supply chains are premier contexts to study TCE and IORs because they focus on an overarching goal to maximize performance while minimizing transaction costs between organizations (Shook et al., 2009).

The Bloomberg SPLC dataset is conducive to further test TCE due to the extensive longitudinal buyer–supplier data with accompanying exposure information. Assuming that each buyer–supplier relationship is accompanied with certain transaction costs, researchers can use Bloomberg SPLC to begin comparing the performance and efficiency outcomes of firms who manage many supplier or buyer relationships with smaller exposures compared to firms with fewer relationships but with larger exposures. Using longitudinal data to look at differences within and between firms may also prove valuable for TCE researchers.

In considering a simplified example, to test TCE using Bloomberg SPLC data, researchers could collect longitudinal supplier data from Apple (and other technology firms) for the past 10 years. Controlling for confounding variables, the study could identify both within and between firms whether a group of consolidated but concentrated suppliers results in greater firm performance than a more diversified approach. Interestingly, under what conditions do we find this tenant of TCE does not hold true in vertical IORs?

Resource-based View and Relational View

The RBV argues that firms can develop strategic resources that allow them to sustain competitive advantages (Barney, 1991). These resources must be valuable, rare, imperfectly imitable, and not substitutable to create such advantages (Barney, 1991). As the RBV evolved, researchers theorized that such resources can be created or obtained through IORs – the latter a core tenet of the relational view (Dyer & Singh, 1998). This perspective focuses on dyad/network routines and process as a unit of analysis to better understand competitive advantages created through IORs. The relational view specifies that increased firm performance is a result of interfirm knowledge-sharing, complementary resource endowments, and effective governance (Dyer & Singh, 1998). Building on this view, one might examine whether firms that transact with one another at greater levels are able to build more relational resources, whether such resources accumulate more on the buyer or supplier side, and whether the competitive environment shapes the nature of resource accumulation. Alternatively, when the collective firms share a similar fate, the relational view suggests that they may form a coalition and lobby governments for changes to laws or regulations that would impact relevant parties. This coalition of firms allows firms to combine their resources, and, in doing so, perhaps creates a "higher order" resource that results in a competitive advantage and increased firm performance for the coalition members.

Researchers have also started recognizing a firm's more extended network as a source of competitive advantage (e.g., Kumar & Zaheer, 2019; Li, de Zubielqui, & O'Connor, 2015). Social networks are systems created by agentic agents (Borgatti, Everett, & Johnson, 2018), and these networks play an important role in determining firms' success. Bloomberg SPLC provides the ability for researchers to answer questions relating to how vertical interorganizational networks can be sources of competitive advantage to a firm. For example, do firms that bridge structural holes – or connect otherwise unconnected firms – in supplier–buyer networks have a competitive advantage that results in increased performance? Or do firms who participate in tight networks of interlinked buyers and suppliers outperform competitors? What type of network characteristics provide firms with competitive advantages? Data from Bloomberg SPLC can be used to answer these types of questions and help us further our understanding of how interfirm relations and networks help cultivate valuable resources.

To test RBV using Bloomberg SPLC data, researchers can take various approaches. First, dyad relations and whether two firms who both share high cost and revenue exposures, but possess key strategic resources, are able to outperform competitors could be examined. In other words, when two firms have a strong buyer–supplier relationship and the requisite resources, does this result in a competitive advantage for the firms? When? What characteristics of the circumstance or relationship lead to competitive advantages?

Next, we can think of the entire supply chain network as a source of competitive advantage. For example, using social network software, we could map out the network of a sample of firms. We could then hypothesize and test what network relationships – or firm locations within the networks – may provide a competitive advantage.

Resource Dependence Theory

This theory focuses on how organizations interact with environments to remove uncertainty and access valuable resources (Pfeffer & Salancik, 1978). First, from an IOR perspective, researchers are interested in how uncertainty can be reduced by interfirm cooperation and the building of joint dependence through IORs. Put differently, RDT asks how firms can create IORs to obtain important resources from the environment that eliminates, in part, some of the power that others hold (Parmigiani & Rivera-Santos, 2011).

Next, in any relationship of two or more actors, RDT posits that social exchange is characterized by both power and dependence (Emerson, 1962). For example, in a buyer–supplier relationship, the buyer may be dependent on the supplier's product. If the supplier is the only firm that makes the product, the supplier has great power over the buyer, since the buyer cannot obtain the resource elsewhere. Thus, the buyer's dependence on the supplier results in the supplier's increased power. However, RDT also states that in unbalanced relationships, actors will strive to minimize the power and dependence gap. Thus, firms will look for strategies to equalize the relationship (e.g., use of alternative resources, political action, coalitions).

By disclosing cost and revenue exposures of buyers and suppliers, Bloomberg SPLC offers new insights into power and dependence relationships (i.e., do suppliers emerge in same industry?). In addition, through the construction of buyer–supplier networks, researchers may be able to identify buyer–supplier coalitions that have combined resources to gain power over other parties (e.g., legislators, competitors). In one of the studies that have leveraged Bloomberg SPLC data, Elking, Paraskevas, Grimm, Corsi, and Steven (2017) examined how supply chain power impacts firm financial performance. This article highlights how Bloomberg SPLC may be used to explore hypotheses related to RDT.

Continuing with Apple as an example, we can see that as of October 2018, Jabil received 28% of its revenue from Apple, and Apple only purchased about 4% of its cost of goods sold from Jabil (see Appendix 2). Hence, it appears that Jabil is more dependent on Apple than Apple is dependent on Jabil in terms of revenues and cost. According to RDT, Jabil would theoretically try to either (1) minimize this dependency on Apple or (2) strive to increase its power in the imbalanced relationship. Historically, do relationships such as this become more stable over time? If not, what are these imbalanced and more dependent firms doing to increase power over the less dependent firm?

Pairing Bloomberg SPLC data with board of director data could also yield a promising study. Does Apple have any board members who are linked with boards of pivotal suppliers or buyers, and is this the reason (i.e., causal mechanism) to increase power in the relationship? Is this more common with firms who receive a large percentage of Apple's revenues?

Stakeholder Theory

This theory claims that the relationships a firm has with its primary and secondary stakeholders (e.g., employees, suppliers, customers, shareholders, media, government) impact the firm's performance (Donaldson & Preston, 1995; Laplume, Sonpar, & Litz, 2008). From a stakeholder theory perspective, a manager's key responsibility is to balance stakeholders' interests, and creating IORs is one mechanism through which to do so. In terms of IORs, stakeholder theory argues that organizations can gain stakeholder support through establishing relationships with other reputable organizations and partners (c.f. Laplume et al., 2008). Through relationships with reputable suppliers and buyers, firms are better able to appeal to stakeholders (Parmigiani & Rivera-Santos, 2011).

We see at least three potential avenues of research using this perspective. The first is to see how, when, and where a firm may highlight its customers and suppliers in order to increase its reputation and status. For example, do firms with powerful and reputable customers and suppliers explicitly communicate these relationships in an attempt to improve stakeholder relationships? Next, researchers can use Bloomberg SPLC to create a natural experiment to explore how a firm's suppliers and buyers change after a negative event. In other words, in attempt to retain legitimacy and reputation, how do other firms respond to a firm whose legitimacy has decreased? An extreme example may be when a firm is exposed of utilizing child labor and inflicting extensive abuse upon its employees, how might their supplier and buyer networks – and the exposure of past relationships – change as a result.

A final suggestion is to use the data to evaluate the supply chains of firms who claim to abide to particular morals or missions – thus appealing to a specific group of stakeholders. In the past, firms' supply chains have been relatively unknown, however, with the data collection efforts of Bloomberg, new relationships are being created on a daily basis. By pairing the Bloomberg SPLC network data with qualitative data from firms' websites, researchers can better understand how organizations may build buyer–supplier networks based on these moral principles, or even how they may hide relationships that do not necessarily align with the standards expected by stakeholders.

Again, let's use Apple as an example. To test notions of stakeholder theory, first, we could compile a list of Apple's suppliers using Bloomberg SPLC. We could then analyze the suppliers' websites and company issued reports for mentions of Apple as a customer of the firms. Are suppliers leveraging this relationship with a well-known firm (Apple) to convey a positive image to stakeholders (e.g., customers, investors)? Next, we can use events to create case studies to understand firms' responses when a supplier acts unethically or against the expectations of the firm and its stakeholders. Suppose one of Apple's key suppliers was involved in a serious scandal in 2012 that was widely publicized by the media. Did Apple decrease its reliance on this supplier or find a new supplier to replace it in order to deflect the potential spillover of stigma and to appease stakeholders? Also, how did this supplier's supplier–buyer network change as a result? Finally, consider firms with strong social missions. Do firms actively promote suppliers in their supply chain that fulfill a similar mission? For example, suppose Apple has a mission of promoting fair labor practices – arguably a strategy to appease its customers. By obtaining a list of Apple's suppliers, one could use content analysis to see which firms also promote fair labor practices and whether Apple (or its suppliers) highlights or hides particular relationships in communications to stakeholders due to their congruence or incongruence with this mission. Also, one could evaluate whether the relationship strength between Apple and its suppliers is related to a matching social mission.

Institutional Theory

Early research in institutional theory focused on the factors that lead to organizations adopting similar structures, strategies, and processes – or how organizations became more isomorphic (e.g., Deephouse, 1996; DiMaggio & Powell, 1983). These scholars looked at how firms sought to increase legitimacy, reputation, and status from appearing to operate like other similar firms. These factors could be coercive (i.e., regulatory), normative (i.e., accepted social values or practice), or mimetic (i.e., copying others of higher status). More recently, however, researchers have been focusing on why and with what consequences organizations exhibit particular arrangements that defy traditional rational explanations (e.g., Harmon, 2018) or when organizations decouple from traditional expectations (Greenwood, Oliver, Sahlin, & Suddaby, 2008).

Similar to stakeholder theory, institutional theory might be leveraged to understand when buyer–supplier networks begin to look, act, and think similarly. For example, in a supply chain where the largest firm has very clear corporate social responsibility goals, do these initiatives trickle up or down the vertical relationships? Do smaller firms engage in normative or mimetic actions to adhere to the behavior of more legitimate firms in their buyer–supplier networks? In other words, do firms feel institutional pressure to conform to other firms' behavior in buyer–supplier networks? To answer these questions, researchers could create a sample of firms that fit a particular criterion. Next, a dataset of these firms' buyer–supplier networks could be constructed utilizing Bloomberg SPLC. Finally, researchers could qualitatively analyze websites, public documents, social media accounts, and other communications of firms within the supply chain to test whether certain norms are institutionalized in buyer–supplier networks.

We could use Bloomberg SPLC to test this theory in multiple ways. For example, if Apple, as a legitimate organization, were to announce an initiative focused on reducing carbon emissions, would Apple's suppliers or buyers also be influenced to adopt this policy? Would other firms in the technology sector follow suit? Moreover, by collecting data from and creating supply chain networks for many firms in an industry, we could test to what extent industries tend to have institutionalized supply chain compositions, and whether or not firms who deviate from this norm have increased or decreased firm performance within an industry.

Social Network Theory

Finally, testing of social network theory can benefit from Bloomberg SPLC data. Most social network research has been conducted using field studies (Borgatti et al., 2018, p. 30) relying on primary data sources (p. 35). These primary data sources include surveys asking individuals about their advice or friendship networks, for example. Other secondary sources of network data include archival records such as business partnerships, voting records, ledger sheets, and accounts of trade (p. 63). More frequently, however, social network researchers are utilizing secondary data sources due to the increased availability of electronic records and data scraping techniques. Some of the sources include bibliometric data, membership data, board member data, electronic communication records, and social media data (p. 35). These sources are valuable because they are often longitudinal in nature, they are easier to collect than conducting surveys, and the data are often more accurate because they highlight actual relationships, and the data are not subject to common method biases.

Historically, obtaining buyer–supplier network information has been difficult. However, Bloomberg's extensive dataset with tens-of-thousands of buyer–

supplier relationships now makes it more feasible. Bloomberg SPLC data can be useful to social network researchers because it can provide information about network qualities (e.g., structural holes, focality, ego networks) and how they impact firm outcomes. In addition, the strength of the relationships can be measured. This results in the ability to test what type of position in a social network is most productive for firms. By utilizing longitudinal data, researchers can also better understand how buyer–supplier networks change over time. For example, how does a negative event between two firms in a network impact other relationships in the network? Finally, while past research has examined interlocking directors among top companies (e.g., Haunschild & Beckman, 1998; Shropshire, 2010; Zona, Gomez-Mejia, & Withers, 2018), researchers have yet to understand how interlocking directors can impact other IORs such as buyer–supplier relationships. Perhaps a social network perspective could also shed light on the mechanisms firms use to decrease power within the supply chain (e.g., create a board link in a buyer–supplier relationship).

Bloomberg SPLC is also a more comprehensive source of data for social network research due to the information regarding relationship size for each partnership. It would be particularly interesting to construct a network of firms in particular industries (e.g., US technology industry) that may share buyers and suppliers. To what extent are suppliers and buyers strategically partnering with specific firms in the industry? What are the predictors and outcomes of particular cost and revenue exposure relationships? How does a firm's position in the network impact firm outcomes? For example, is Jabil – a supplier of Apple that receives 28% of its revenue from the relationship – positioned differently than Intel, a supplier of Apple that only receives about 6% of its revenue from Apple, in the broader supply chain (Appendix 2)? Are weak or strong ties more valuable in a supplier–buyer network? Is centrality a key to firm success in supplier–buyer relationships? These and many more questions can be addressed by those interested in using Bloomberg SPLC to research social network theories.

LIMITATIONS

Although Bloomberg SPLC is a useful source of IOR data, we acknowledge several limitations that should be considered when using the dataset. First, as previously mentioned, Bloomberg SPLC compiles a variety of sources and uses an algorithm to estimate the revenue and cost exposures for relationships with an undisclosed size. While these estimates are expected to approximate the actual relationship, with the current information provided by Bloomberg SPLC, we do not know the precise accuracy of these estimates.[3] Next, Bloomberg SPLC categorizes a supplier by placing it into an accounting-type category

[3]Bloomberg has mentioned disclosing statistical confidence intervals for estimated relationships, but this has not yet been implemented into Bloomberg SPLC (Davenport, 2011).

(i.e., COGS, CAPEX, R&D, SG&A). Suppliers who are diversified and provide goods or services that transcend multiple cost categories may only be represented in one category in the data. We recommend minimizing this limitation by including firm diversification as a control variable in research studies. Finally, when searching historical relationships using the "as of date" feature in Bloomberg SPLC, researchers should be aware that the data that populate are the most recent estimate Bloomberg SPLC has for each relationship. As mentioned, Bloomberg SPLC is constantly updating relationship information, and each relationship presented may not coincide with the entered date. For example, although 12/18/2018 was entered in the "as of" section in Bloomberg SPLC, the Apple and Intel relationship was last updated on 4/4/18 and more accurately represents the relationship exposures as of that date (Appendix 2). Hence, researchers should be aware of this nuance in the data to ensure the data accurately capture and represent the appropriate information required for a study.

CONCLUSION

There is no doubt that firms are becoming increasingly reliant on IORs for their survival and prosperity. These relationships come in a variety of shapes and sizes, such as strategic alliances, joint ventures, buyer–supplier agreements, franchising, cross-sector partnerships, networks, trade associations, interlocking directorates, and coalitions. However, while we know IORs are important in today's business landscape and that firms' fates are often linked with other firms (Parmigiani & Rivera-Santos, 2011), researchers currently have relatively few secondary data sources to empirically test hypotheses related to IORs. In this chapter, we have highlighted a relatively new data source, Bloomberg SPLC, as a potential trove of information that can be used to test theories such as TCE, RBV, resource dependency theory, stakeholder theory, institutional theory, and social network theory. Bloomberg SPLC reports data at various levels from dyadic relationships to complete industry buyer–supplier networks, and also quantifies these relationships – providing exposure and risk information across IORs. Given this, the Bloomberg SPLC opens up a variety of exciting research opportunities, especially when paired with other datasets, and we encourage researchers studying IORs to consider Bloomberg SPLC as a potential resource now and in the future.

REFERENCES

Ahuja, G., Lampert, C. M., & Tandon, V. (2008). 1 moving beyond Schumpeter: Management research on the determinants of technological innovation. *The Academy of Management Annals*, *2*(1), 1–98.

Aleman, E. A. (2018). Release No. 33-10532; 34-83875; IC-33203; File No. S7-15-16 (United States, Securities and Exchange Commission). Retrieved from https://www.sec.gov/rules/final/2018/33-10532.pdf. Accessed on August 2, 2019.

Asgari, N., Singh, K., & Mitchell, W. (2017). Alliance portfolio reconfiguration following a technological discontinuity. *Strategic Management Journal*, *38*(5), 1062–1081.
Bakker, R. M. (2016). Stepping in and stepping out: Strategic alliance partner reconfiguration and the unplanned termination of complex projects. *Strategic Management Journal*, *37*(9), 1919–1941.
Barney, J. (1991). Firm resources and sustained competitive advantage. *Journal of Management*, *17*(1), 99–120.
Barringer, B. R., & Harrison, J. S. (2000). Walking a tightrope: Creating value through interorganizational relationships. *Journal of Management*, *26*(3), 367–403.
Blevins, D. P., & Ragozzino, R. (2018). An examination of the effects of venture capitalists on the alliance formation activity of entrepreneurial firms. *Strategic Management Journal*, *39*(7), 2075–2091.
Bloomberg Professional Services. (2017). *As seen on 60 minutes. August 25, 2017*. Retrieved from https://www.bloomberg.com/professional/blog/seen-60-minutes/. Accessed on August 3, 2019.
Borgatti, S. P., Everett, M. G., & Johnson, J. C. (2018). *Analyzing social networks*. London: SAGE Publications.
Bos, B., Faems, D., & Noseleit, F. (2017). Alliance concentration in multinational companies: Examining alliance portfolios, firm structure, and firm performance. *Strategic Management Journal*, *38*(11), 2298–2309.
Brouthers, K. D., & Hennart, J. F. (2007). Boundaries of the firm: Insights from international entry mode research. *Journal of Management*, *33*(3), 395–425.
Burt, R. S. (1980). Cooptive corporate actor networks: A reconsideration of interlocking directorates involving American manufacturing. *Administrative Science Quarterly*, *25*(4), 557–582.
Combs, J. G., & Ketchen, Jr. D. J. (2003). Why do firms use franchising as an entrepreneurial strategy?: A meta-analysis. *Journal of Management*, *29*(3), 443–465.
Connelly, B. L., Crook, T. R., Combs, J. G., Ketchen, Jr. D. J., & Aguinis, H. (2018). Competence- and integrity-based trust in interorganizational relationships: Which matters more? *Journal of Management*, *44*(3), 919–945.
Crook, T. R., & Combs, J. G. (2007). Sources and consequences of bargaining power in supply chains. *Journal of Operations Management*, *25*(2), 546–555.
Crook, T. R., Combs, J. G., Ketchen, Jr. D. J., & Aguinis, H. (2013). Organizing around transaction costs: What have we learned and where do we go from here? *Academy of Management Perspectives*, *27*, 63–79.
Dacin, M. T., Oliver, C., & Roy, J. P. (2007). The legitimacy of strategic alliances: An institutional perspective. *Strategic Management Journal*, *28*(2), 169–187.
Davenport, R. (2011). Supply chain on Bloomberg. Emory Libraries & Information Technology. Retrieved from https://business.library.emory.edu/documents/faq-handouts/bloomberg-splc.pdf
Deephouse, D. L. (1996). Does isomorphism legitimate? *Academy of Management Journal*, *39*(4), 1024–1039.
Devarakonda, S. V., & Reuer, J. J. (2018). Knowledge sharing and safeguarding in R&D collaborations: The role of steering committees in biotechnology alliances. *Strategic Management Journal*, *39*(7), 1912–1934.
DiMaggio, P. J., & Powell, W. W. (1983). The iron cage revisited: Institutional isomorphism and collective rationality in organizational fields. *American Sociological Review*, *48*(2), 147–160.
Donaldson, T., & Preston, L. E. (1995). The stakeholder theory of the corporation: Concepts, evidence, and implications. *Academy of Management Review*, *20*(1), 65–91.
Dyer, J. H., & Singh, H. (1998). The relational view: Cooperative strategy and sources of interorganizational competitive advantage. *Academy of Management Review*, *23*(4), 660–679.
Elking, I., Paraskevas, J. P., Grimm, C., Corsi, T., & Steven, A. (2017). Financial dependence, lean inventory strategy, and firm performance. *Journal of Supply Chain Management*, *53*(2), 22–38.

Elston, T., MacCarthaigh, M., & Verhoest, K. (2018). Collaborative cost-cutting: Productive efficiency as an interdependency between public organizations. *Public Management Review*, *20*(12), 1815–1835.

Emerson, R. M. (1962). Power-dependence relations. *American Sociological Review*, *27*(1), 31–41.

Ghosh, A., Ranganathan, R., & Rosenkopf, L. (2016). The impact of context and model choice on the determinants of strategic alliance formation: Evidence from a staged replication study. *Strategic Management Journal*, *37*(11), 2204–2221.

Golicic, S. L., Foggin, J. H., & Mentzer, J. T. (2003). Relationship magnitude and its role in interorganizational relationship structure. *Journal of Business Logistics*, *24*(1), 57–75.

Greenwood, R., Oliver, C., Sahlin, K., & Suddaby, R. (2008). *The SAGE handbook of organizational institutionalism*. Los Angeles, CA: SAGE Publications.

Handfield, R. (2012). The power of Bloomberg in creating supply chain intelligence and risk alerts. NC State University. Retrieved from https://scm.ncsu.edu/scm-articles/article/the-power-of-bloomberg-in-creating-supply-chain-intelligence-and-risk-alerts. Accessed on August 2, 2019.

Harmon, D. J. (2018). When the Fed speaks: Arguments, emotions, and the microfoundations of institutions. *Administrative Science Quarterly*, *64*(3), 542–575.

Haunschild, P. R., & Beckman, C. M. (1998). When do interlocks matter?: Alternate sources of information and interlock influence. *Administrative Science Quarterly*, *43*(4), 815–844.

Help Page Supply Chain Analysis (SPLC). (2018). Bloomberg. Retrieved from Bloomberg terminal, 21–22. Accessed on October 23, 2018.

Hoang, H., & Antoncic, B. (2003). Network-based research in entrepreneurship: A critical review. *Journal of Business Venturing*, *18*(2), 165–187.

Holcomb, T. R., & Hitt, M. A. (2007). Toward a model of strategic outsourcing. *Journal of Operations Management*, *25*(2), 464–481.

Howard, M. D., Withers, M. C., Carnes, C. M., & Hillman, A. J. (2016). Friends or strangers? It all depends on context: A replication and extension of Beckman, Haunschild, and Phillips (2004). *Strategic Management Journal*, *37*(11), 2222–2234.

Jamali, D., & Keshishian, T. (2009). Uneasy alliances: Lessons learned from partnerships between businesses and NGOs in the context of CSR. *Journal of Business Ethics*, *84*(2), 277–295.

Jiang, H., Xia, J., Cannella, A. A., & Xiao, T. (2018). Do ongoing networks block out new friends? Reconciling the embeddedness constraint dilemma on new alliance partner addition. *Strategic Management Journal*, *39*(1), 217–241.

Kim, J. Y., Howard, M., Cox Pahnke, E., & Boeker, W. (2016). Understanding network formation in strategy research: Exponential random graph models. *Strategic Management Journal*, *37*(1), 22–44.

Kim, Y., & Choi, T. Y. (2018). Tie strength and value creation in the buyer-supplier context: A U-shaped relation moderated by dependence asymmetry. *Journal of Management*, *44*(3), 1029–1064.

Kim, Y. H., & Davis, G. F. (2016). Challenges for global supply chain sustainability: Evidence from conflict minerals reports. *Academy of Management Journal*, *59*(6), 1896–1916.

Kumar, P., & Zaheer, A. (2019). Ego-network stability and innovation in alliances. *Academy of Management Journal*, *62*(3), 691–716. doi:10.5465/amj.2016.0819

Laplume, A. O., Sonpar, K., & Litz, R. A. (2008). Stakeholder theory: Reviewing a theory that moves us. *Journal of Management*, *34*, 1152–1189.

Li, H., de Zubielqui, G. C., & O'Connor, A. (2015). Entrepreneurial networking capacity of cluster firms: A social network perspective on how shared resources enhance firm performance. *Small Business Economics*, *45*(3), 523–541.

Man Zhang, C., & Greve, H. R. (2019). Dominant coalitions directing acquisitions: Different decision makers, different decisions. *Academy of Management Journal*, *62*(1), 44–65.

March, J. G. (1991). Exploration and exploitation in organizational learning. *Organization Science*, *2*(1), 71–87.

Mawdsley, J. K., & Somaya, D. (2018). Demand-side strategy, relational advantage, and partner-driven corporate scope: The case for client-led diversification. *Strategic Management Journal, 39*(7), 1834–1859.
McCutcheon, D., & Stuart, F. I. (2000). Issues in the choice of supplier alliance partners. *Journal of Operations Management, 18*(3), 279–301.
Mindruta, D., Moeen, M., & Agarwal, R. (2016). A two-sided matching approach for partner selection and assessing complementarities in partners' attributes in inter-firm alliances. *Strategic Management Journal, 37*(1), 206–231.
Mizruchi, M. S. (1996). What do interlocks do? An analysis, critique, and assessment of research on interlocking directorates. *Annual Review of Sociology, 22*(1), 271–298.
Munyon, T. P., Jenkins, M. T., Crook, T. R., Edwards, J., & Harvey, N. P. (2019). Consequential cognition: Exploring how attribution theory sheds new light on the firm-level consequences of product recalls. *Journal of Organizational Behavior, 40*(5), 587–602.
Oliveira, N., & Lumineau, F. (2019). The dark side of interorganizational relationships: An integrative review and research agenda. *Journal of Management, 45*(1), 231–261.
Oliver, C. (1990). Determinants of interorganizational relationships: Integration and future directions. *Academy of Management Review, 15*(2), 241–265.
Park, N. K., & Mezias, J. M. (2005). Before and after the technology sector crash: The effect of environmental munificence on stock market response to alliances of e-commerce firms. *Strategic Management Journal, 26*(11), 987–1007.
Parmigiani, A., & Rivera-Santos, M. (2011). Clearing a path through the forest: A meta-review of interorganizational relationships. *Journal of Management, 37*(4), 1108–1136.
Pfeffer, J., & Salancik, G. R. (1978). *The external control of organizations: A resource dependence perspective*. New York, NY: Harper & Row.
Polzer, J. T., Mannix, E. A., & Neale, M. A. (1998). Interest alignment and coalitions in multiparty negotiation. *Academy of Management Journal, 41*(1), 42–54.
Provan, K. G., Fish, A., & Sydow, J. (2007). Interorganizational networks at the network level: A review of the empirical literature on whole networks. *Journal of Management, 33*(3), 479–516.
Sauerwald, S., Lin, Z., & Peng, M. W. (2016). Board social capital and excess CEO returns. *Strategic Management Journal, 37*(3), 498–520.
Scott, W. R. (1987). *Organizations: Rational, natural, and open systems*. Englewood Cliffs, NJ: Prentice-Hall.
Selsky, J. W., & Parker, B. (2005). Cross-sector partnerships to address social issues: Challenges to theory and practice. *Journal of Management, 31*(6), 849–873.
Shook, C. L., Adams, G. L., Ketchen, Jr. D. J., & Craighead, C. W. (2009). Towards a "theoretical toolbox" for strategic sourcing. *Supply Chain Management: International Journal, 14*(1), 3–10.
Shropshire, C. (2010). The role of the interlocking director and board receptivity in the diffusion of practices. *Academy of Management Review, 35*(2), 246–264.
Steven, A. B., Dong, Y., & Corsi, T. (2014). Global sourcing and quality recalls: An empirical study of outsourcing supplier concentration-product recalls linkages. *Journal of Operations Management, 32*, 241–253.
Villena, V. H., Choi, T. Y., & Revilla, E. (2019). Revisiting interorganizational trust: Is more always better or could more be worse? *Journal of Management, 45*(2), 752–785.
Wassmer, U. (2010). Alliance portfolios: A review and research agenda. *Journal of Management, 36*(1), 141–171.
Williamson, O. E. (1985). *The economic institutions of capitalism: Firms, markets, relational contracting*. New York, NY: Free Press.
Zald, M. N. (1978). On the social control of industries. *Social Forces, 57*(1), 79–102.
Zimmermann, F., & Foerstl, K. (2014). A meta-analysis of the "purchasing and supply management practice–performance link". *Journal of Supply Chain Management, 50*(3), 37–54.
Zona, F., Gomez-Mejia, L. R., & Withers, M. C. (2018). Board interlocks and firm performance: Toward a combined agency–resource dependence perspective. *Journal of Management, 44*(2), 589–618.

APPENDIX 1

Bloomberg SPLC Chart View

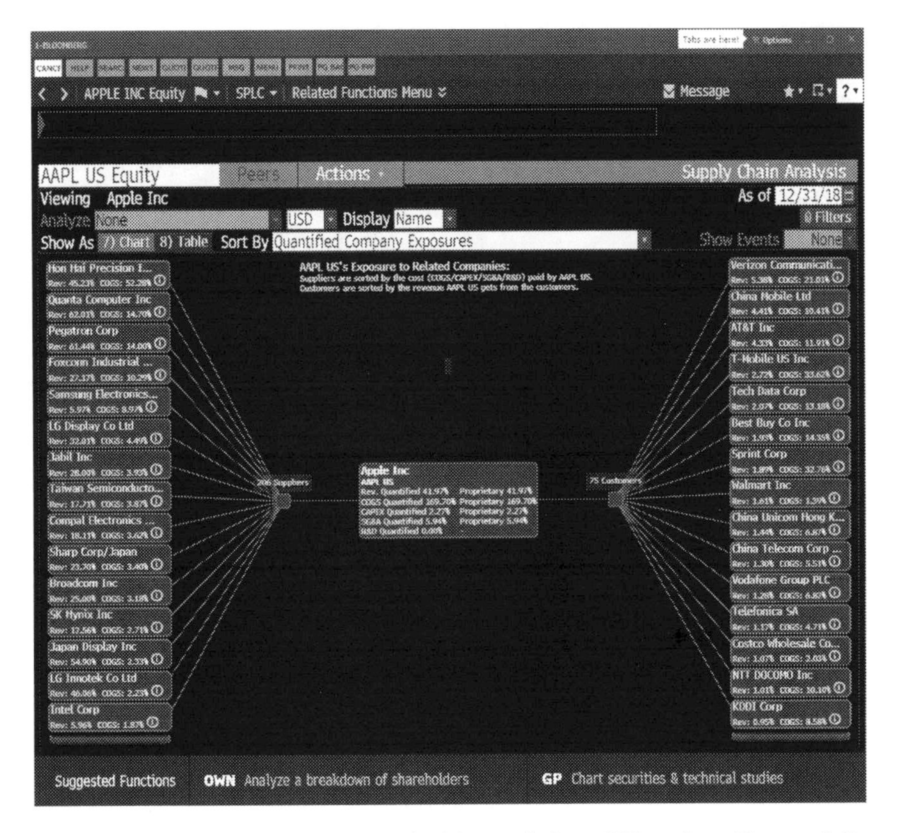

Source: Bloomberg Finance L.P. Used with permission of Bloomberg Finance L.P.

APPENDIX 2

Bloomberg SPLC Table View

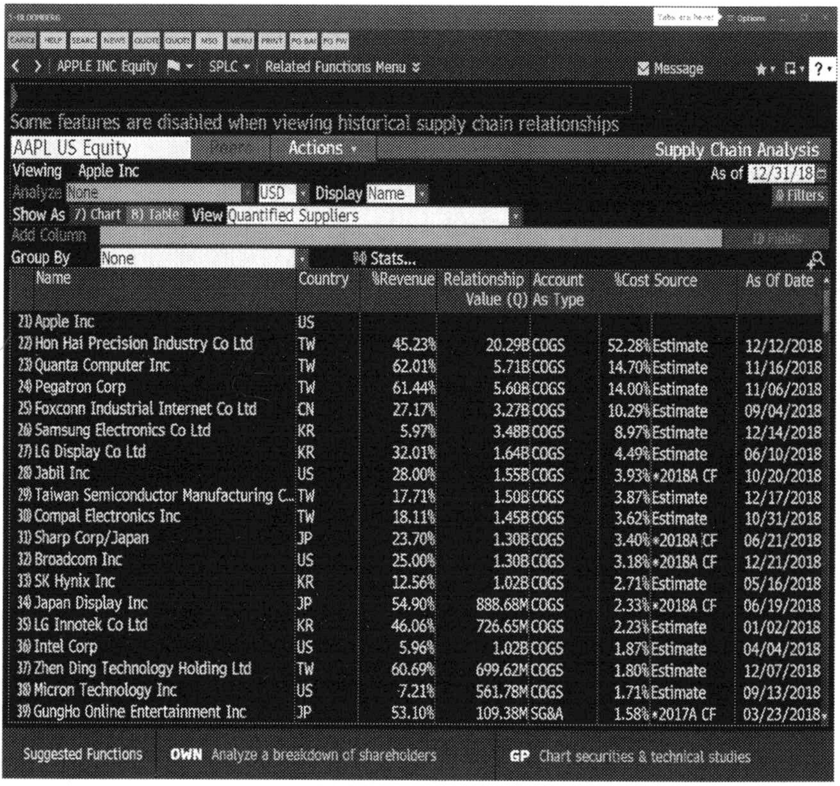

< > APPLE INC Equity SPLC Related Functions Menu Message

Some features are disabled when viewing historical supply chain relationships

AAPL US Equity Actions Supply Chain Analysis

Viewing Apple Inc As of 12/31/18

Analyze None USD Display Name Filters

Show As 7) Chart 8) Table View Quantified Suppliers

Add Column

Group By None 94) Stats...

Name	Country	%Revenue	Relationship Value (Q)	Account As Type	%Cost	Source	As Of Date
21) Apple Inc	US						
22) Hon Hai Precision Industry Co Ltd	TW	45.23%	20.29B	COGS	52.28%	Estimate	12/12/2018
23) Quanta Computer Inc	TW	62.01%	5.71B	COGS	14.70%	Estimate	11/16/2018
24) Pegatron Corp	TW	61.44%	5.60B	COGS	14.00%	Estimate	11/06/2018
25) Foxconn Industrial Internet Co Ltd	CN	27.17%	3.27B	COGS	10.29%	Estimate	09/04/2018
26) Samsung Electronics Co Ltd	KR	5.97%	3.48B	COGS	8.97%	Estimate	12/14/2018
27) LG Display Co Ltd	KR	32.01%	1.64B	COGS	4.49%	Estimate	06/10/2018
28) Jabil Inc	US	28.00%	1.55B	COGS	3.93%	*2018A CF	10/20/2018
29) Taiwan Semiconductor Manufacturing C...	TW	17.71%	1.50B	COGS	3.87%	Estimate	12/17/2018
30) Compal Electronics Inc	TW	18.11%	1.45B	COGS	3.62%	Estimate	10/31/2018
31) Sharp Corp/Japan	JP	23.70%	1.30B	COGS	3.40%	*2018A CF	06/21/2018
32) Broadcom Inc	US	25.00%	1.30B	COGS	3.18%	*2018A CF	12/21/2018
33) SK Hynix Inc	KR	12.56%	1.02B	COGS	2.71%	Estimate	05/16/2018
34) Japan Display Inc	JP	54.90%	888.68M	COGS	2.33%	*2018A CF	06/19/2018
35) LG Innotek Co Ltd	KR	46.06%	726.65M	COGS	2.23%	Estimate	01/02/2018
36) Intel Corp	US	5.96%	1.02B	COGS	1.87%	Estimate	04/04/2018
37) Zhen Ding Technology Holding Ltd	TW	60.69%	699.62M	COGS	1.80%	Estimate	12/07/2018
38) Micron Technology Inc	US	7.21%	561.78M	COGS	1.71%	Estimate	09/13/2018
39) GungHo Online Entertainment Inc	JP	53.10%	109.38M	SG&A	1.58%	*2017A CF	03/23/2018

Suggested Functions OWN Analyze a breakdown of shareholders GP Chart securities & technical studies

Source: Bloomberg Finance L.P. Used with permission of Bloomberg Finance L.P.

A GUIDE FOR CONDUCTING CURVILINEAR META-ANALYSES

Jeremy D. Mackey, Charn P. McAllister, Liam P. Maher and Gang Wang

ABSTRACT

Recently, there has been an increase in the number and type of studies in the organizational sciences that examine curvilinear relationships. These studies are important because some relationships have context-specific inflection points that alter their magnitude and/or direction. Although some scholars have utilized basic techniques to make meta-analytic inferences about curvilinear effects with the limited information available about them, there is still a tremendous opportunity to advance our knowledge by utilizing rigorous techniques to meta-analytically examine curvilinear effects. In a recent study, we used a novel meta-analytic approach in an effort to comprehensively examine curvilinear relationships between destructive leadership and followers' workplace outcomes. The purpose of this chapter is to provide an actionable guide for conducting curvilinear meta-analyses by describing the meta-analytic techniques we used in our recent study. Our contributions include a detailed guide for conducting curvilinear meta-analyses, the useful context we provide to facilitate its implementation, and our identification of opportunities for scholars to leverage our technique in future studies to generate nuanced knowledge that can advance their fields.

Keywords: Meta-analysis; curvilinear; nonlinear; guide; method; meta

Advancing Methodological Thought and Practice
Research Methodology in Strategy and Management, Volume 12, 101–115

ISSN: 1479-8387/doi:10.1108/S1479-838720200000012018

A GUIDE FOR CONDUCTING CURVILINEAR META-ANALYSES

Perhaps too much of everything is as bad as too little. ~ Edna Ferber.

Although the notion of having too little of something is a well-accepted truism, the notion that having "too much of a good thing" is becoming more and more accepted in business research. Accordingly, there has been an uptick in research that examines whether there are optimal points that exist where there is neither too little nor too much of something; this includes the ever-elusive search for inflection points at which seemingly universal positives (e.g., job satisfaction) and negatives (e.g., abusive supervision) suddenly start to produce counterintuitive results. Much of this research has emphasized improving our understanding of the extent to which linear versus curvilinear terms predict important outcomes. Curvilinear relationships exist when context-specific inflection points alter the magnitude and/or direction of relationships (e.g., Pierce & Aguinis, 2013). Some scholars have used theory (Shepherd & Suddaby, 2017) to explain the effects of curvilinear terms (e.g., Barnett & Salomon, 2006; Harris, Kacmar, & Witt, 2005). Other scholars have examined curvilinear terms as control variables (e.g., Mackey, Ellen III, Hochwarter, & Ferris, 2013) because they can overlap with interaction effects (Cortina, 1993) and/or provide useful supplementary information for regression analyses (Edwards, 2008). Regardless of the purpose of their use, it is clear that examining curvilinear terms can facilitate the creation of nuanced knowledge that can advance numerous academic literatures.

However, only a small percentage of organizational studies actually report results that include curvilinear terms. Even fewer studies report curvilinear terms in correlation matrices. The aforementioned trends make it tremendously difficult to accurately determine the replicability and generalizability of curvilinear effects across studies. Although some scholars have utilized basic techniques to conduct meta-analyses with the limited information available about curvilinear effects, there is still a tremendous opportunity to advance our knowledge by utilizing rigorous techniques to meta-analytically examine curvilinear effects. Scholars are beginning to illuminate the valuable insights that curvilinear meta-analyses have to offer. For example, we used advanced meta-analytic techniques for comprehensively examining linear and curvilinear relationships between destructive leadership and followers' workplace outcomes in a recent study (i.e., Mackey, McAllister, Maher, & Wang, 2019). The primary benefit of our approach was that it enabled us to incorporate results from raw data instead of make inferences based on categorical information (e.g., study design features) that may or may not accurately portray curvilinear effects.

The purpose of this chapter is to provide a guide for conducting curvilinear meta-analyses by describing the meta-analytic techniques we used in our recent study. Throughout this chapter, we provide an overview of meta-analysis in general, describe some of the prior techniques for meta-analytically examining curvilinear relationships, illuminate our process for conducting meta-analyses that examine curvilinear effects, detail some of the challenges we encountered with our approach, and identify some actionable opportunities for scholars to leverage our techniques in future research. Overall, our contributions stem from the actionable guide for conducting curvilinear meta-analyses and the useful context for its implementation that we provide.

OVERVIEW OF META-ANALYTIC TECHNIQUES

Meta-analysis (Glass, 1976) was originally developed to address the need to combine findings across studies that examined the same relationship (Hunter & Schmidt, 2004). Many early meta-analytic techniques emphasized conducting systematic empirical reviews that compared significant versus nonsignificant results (Combs, Crook, & Rauch, 2019). However, contemporary approaches heavily emphasize examining the magnitude of correlation/effect sizes instead of the statistical significance of relationships across studies (Hunter & Schmidt, 2004). In particular, current meta-analytic approaches emphasize estimating the magnitude and variation of effect sizes across studies (Carlson & Ji, 2011).

Meta-analyses are important because they provide the basis for theory development by integrating findings across "studies to reveal the simpler patterns of relationships that underlie research literatures" (Hunter & Schmidt, 2004, p. 17). Scholars even argue that "meta-analysis has become essential in the evolution of knowledge" (Combs et al., 2019, p. 1) due to the massive impact it has on the development of academic literatures (Aguinis, Dalton, Bosco, Pierce, & Dalton, 2011). Although "the goal in any science is the production of cumulative knowledge" (Hunter & Schmidt, 2004, p. 17), there are often conflicting findings across studies within research literatures due to a lack of consensus about the precise magnitude of relationships across study contexts (Yu, Downes, Carter, & O'Boyle, 2016). These differences mean it is imperative for researchers to conduct meta-analyses so we can create a robust understanding of the direction, magnitude, and variability (i.e., heterogeneity) of observed effects across studies. Meta-analyses also enable us to generate knowledge about how measurement artifacts (e.g., unreliability of measures) and sampling error (e.g., limitations that stem from using only the data that are available to us) affect our results, as well as identify outlier and/or context-specific findings within literatures.

Best practice recommendations and the standards for conducting meta-analyses continue to improve. In recent years, scholars have emphasized carefully standardizing judgment calls (Aguinis et al., 2011), assessing power (Cafri, Kromrey, & Brannick, 2010), ensuring sample independence by carefully examining studies with duplicate effect sizes (Wood, 2008), addressing publication bias (Dalton, Aguinis, Dalton, Bosco, & Pierce, 2012), evaluating outliers (Aguinis, Gottfredson, & Joo, 2013), assessing range restriction (Hunter, Schmidt, & Le, 2006), justifying analytical approaches (Borenstein, Hedges, Higgins, & Rothstein, 2010), and accounting for effect size heterogeneity (Yu et al., 2016). There has even been an emphasis on contextualizing meta-analytic findings appropriately for scholars (Desimone, Köhler, & Schoen, 2019) and practitioners (Ones, Viswesvaran, & Schmidt, 2017) alike. The results of these efforts have been an improved understanding about the utility of meta-analyses (Aguinis, Pierce, Bosco, Dalton, & Dalton, 2011), greater awareness regarding how to implement best practices recommendations (Geyskens, Krishnan, Steenkamp, & Cunha, 2009), and an emphasis on following high quality meta-analytic reporting standards (Kepes, McDaniel, Brannick, & Banks, 2013). The other consequence of these advancements is an emphasis on theory building and testing in meta-analyses. Most meta-analyses to date examine linear relationships, so the next step of examining curvilinear relationships is needed. Thus, now is an excellent time to leverage curvilinear meta-analyses to generate novel insights.

TRADITIONAL TECHNIQUES FOR EXAMINING CURVILINEAR RELATIONSHIPS

Although great strides have been made to enhance the rigor (Combs et al., 2019) and replicability (Aytug, Rothstein, Zhou, & Kern, 2012) of meta-analyses, there still have been very few meta-analyses that examine curvilinear effects. However, there are a few notable exceptions. For example, Palich, Cardinal, and Miller (2000) used a range restriction technique to meta-analytically examine curvilinearity in the relationship between firm diversification and performance. Essentially, Palich et al. created categories within their meta-analytic dataset so they could conduct subgroup moderator analyses (Gonzalez-Mulé & Aguinis, 2018) that would enable them to examine whether or not there was evidence of asymptotic U-shaped or inverted U-shaped relationships across different levels of firm diversification. Other studies (e.g., Jin et al., 2017; Mackey, Roth, Van Iddekinge, & McFarland, 2019) have since used subgroup moderator analyses to examine curvilinear relationships that were subject to floor and/or ceiling effects (Wang, Zhang, McArdle, & Salthouse, 2008). These studies used contemporary meta-analytic techniques to generate novel insight using the information that was available in primary studies. Our approach to conducting curvilinear meta-analysis goes a step further by requesting previously unreported results from primary study authors' raw data instead of relying on subgroup moderator analyses to make inferences about curvilinearity.

OUR GUIDE FOR CONDUCTING CURVILINEAR META-ANALYSES

Below, we describe our meta-analytic technique for conducting meta-analyses that examine curvilinear effects. The first part of our process stems from following contemporary techniques for conducting linear meta-analyses. Table 1 provides a broad overview of the basic steps we follow for conducting linear and curvilinear meta-analyses. Rather than review our techniques for linear meta-analyses here, we defer to other resources that provide best practice recommendations (e.g., Geyskens et al., 2009; Gonzalez-Mulé & Aguinis, 2018) for utilizing the meta-analysis reporting standards (e.g., Kepes et al., 2013). We recommend reviewing Mackey, McAllister, Ellen III, and Carson (in press) for extensive details regarding how we conduct and report linear meta-analyses. We also recommend reviewing the online supplemental file for Mackey et al. (in press) because it shows how we report our meta-analytic data. Overall, our approach emphasizes that our goal while conducting linear and curvilinear meta-analyses is to be as explicit and transparent as possible (Aguinis, Ramani, & Alabdduljader,

Table 1. Recommendations for Conducting Curvilinear Meta-analyses.

Linear Meta-analysis (see Mackey, McAllister, Ellen III, and Carson, in press)

1. Determine the specific relationship to study.
2. Determine the inclusion criteria for your meta-analysis.
3. Search for studies that meet the inclusion criteria.
4. Finalize the list of studies to include.
5. Create a coding form that lists each study to code and includes columns for sample size (i.e., n), the correlation for the focal relationship (e.g., r), reliability estimates (e.g., α), and any relevant study design characteristics.
6. Two independent coders enter information into the coding form.
7. Compare information across coders, address all discrepancies by referring to the original studies, and calculate interrater agreement for the linear coding information.

Curvilinear Meta-analysis (see Mackey, McAllister, et al., 2019)

8. Create custom SPSS syntax files for authors of each primary study. The syntax files should enable the primary study authors to quickly create new variable names, composites (when necessary), and curvilinear variables (i.e., variable^2), as well as descriptive statistics and correlation matrices.
9. E-mail *all* of the coauthors of each primary study (please see Appendix 2). Request a correlation matrix with linear and curvilinear terms, as well as descriptive statistics. Include a custom syntax file to facilitate the process (please see Appendix 2).
10. Compile a summary document with all of the responses from primary study authors. We recommend creating the reference section, then copy-pasting all of the results primary study authors send below the corresponding reference.
11. Send a follow-up e-mail to primary study authors who did not respond to the initial request for data.
12. Two independent coders enter curvilinear correlation information into the coding form.
13. Compare coders' responses, address all discrepancies by referring to the summary document with data from primary authors, and calculate interrater agreement for curvilinear coding information.
14. Examine the data for outliers (e.g., visually inspect the data, create funnel plots).
15. Conduct the linear meta-analyses.
16. Use the transformation in Appendix E of Mackey et al. (in press) to correct the obtained curvilinear correlations for using uncentered linear terms when creating the curvilinear terms.
17. Create reliability information for each of the curvilinear terms by squaring the value of the reliability estimate of the corresponding linear term (Dimitruk et al., 2007, Eq. (12): $\alpha_{IV}{}^2$).
18. Conduct the meta-analyses necessary to create correlation tables for the curvilinear analyses (i.e., $\rho_{IV\ \&\ DV}$, $\rho_{IV\ \&\ IV^2}$, and $\rho_{IV^2\ \&\ DV}$).
19. Create input correlation tables with the population correlation estimates (i.e., $\rho_{IV\ \&\ DV}$, $\rho_{IV\ \&\ IV^2}$, and $\rho_{IV^2\ \&\ DV}$).

2018) so that our meta-analytic dataset and results are replicable (Aytug et al., 2012). Please see Table 1 for an overview of the key steps we describe throughout the following sections.

Literature Search

Conducting thorough literature searches is especially important for curvilinear meta-analyses because large sample sizes are needed to adequately examine curvilinear effects. Obtaining a large sample size can be particularly difficult for curvilinear meta-analyses because you likely will not receive data from the primary study authors of every study you find for possible inclusion. Thus, precautions should be taken to find both published and unpublished studies (Kepes, Banks, McDaniel, & Whetzel, 2012; Rothstein, Sutton, & Borenstein, 2005).

It is likely that you will have to contact authors of almost every primary study you identify because curvilinear terms typically are not included in correlation matrices. We created a template e-mail that included a brief description of our study (about one sentence) and a request for supplementary information from the study authors (see Appendix 1). Specifically, we requested a correlation matrix and descriptive statistics for the variables we wanted to include in our study. We included the reference for the primary study we identified and a list of variables we wanted to include in our study (e.g., abusive supervision, abusive supervision^2, task performance, task performance^2). Finally, we included any additional information that would be helpful, thanked the recipients for considering our request, and attached a custom SPSS syntax file (see Appendix 2) to each e-mail that would enable the recipients to easily create the variables and run the requested analyses (i.e., correlation matrix and descriptive statistics). We recommend trying to include as many of the primary study authors on the e-mail request as possible because it seemed to increase our response rate. We sent follow-up e-mails to authors who did not respond within one month. Overall, about 59% of authors responded to our request with complete and useable data.

Coding Correlations

Overall, the goals when coding for curvilinear meta-analyses are similar to the goals when coding for linear meta-analyses: create low-inference coding procedures that limit the need for subjective judgment calls (Aguinis et al., 2011). Extensive information about our linear coding system can be found in Mackey et al. (in press). For the curvilinear portion of our coding form, we created columns in the coding form we utilized in Microsoft Excel so coders could record the following key information: correlations, means, and standard deviations. We recorded three main correlations for each study: (1) the correlation (r) between the linear independent variable (IV) and linear dependent variable (DV) (i.e., $r_{IV\ \&\ DV}$), (2) the correlation between the linear IV and the curvilinear IV (i.e., IV^2; $r_{IV\ \&\ IV^2}$), and (3) the correlation between the curvilinear IV and the DV (i.e., $r_{IV^2\ \&\ DV}$). We also recorded the means and standard deviations for the IV, IV^2, and DV. Then, we created a column for the internal consistency (i.e., alpha; α) for the IV^2 term, which was simply the squared value of the IV's reliability estimate (i.e., α_{IV}^2; Dimitruk, Schermelleh-Engel, Kelava, & Moosbrugger, 2007, Eq. 12).

Transforming Correlations

Next, we transformed the correlations between the linear IV and the curvilinear IV (i.e., IV^2) term because linear and curvilinear terms tend to be very highly correlated, which generates concerns about multicollinearity (Disatnik & Sivan, 2016; Shieh, 2011). The transformation alters the correlations to reflect the values that would have been obtained if the primary study authors mean centered the linear term prior to creating the curvilinear term. We recommend transforming the correlations instead of requesting mean-centered correlations from study authors so you can ensure that the transformations were made systematically across studies. We recommend creating a Microsoft Excel file that automatically generates corrected correlations when you enter the key correlation and descriptive statistics information into highlighted cells.

Creating Meta-analytic Correlation Matrices

Many of the supplementary analyses that examine linear versus curvilinear relationships require the use of meta-analytic correlation matrices (Roth, Switzer, Van Iddekinge, & Oh, 2011). We describe our process for creating the meta-analytic correlation matrices below. First, it is necessary to run meta-analyses for each individual correlation that will be reported in the correlation matrix (i.e., $\rho_{IV\ \&\ DV}$, $\rho_{IV\ \&\ IV^2}$, and $\rho_{IV^2\ \&\ DV}$). We recommend using the population correlations (i.e., ρ) that are corrected for measurement and sampling error to create a correlation matrix. The sample size likely will vary according to the type of analysis conducted, but the harmonic mean (i.e., $((n_1+n_2+n_3)/((1/n_1) + (1/n_2) + (1/n_3))))$ provides a useful reference point. Creating a meta-analytic correlation matrix is foundational to implementing many different analyses, including regression-based procedures (Viswesvaran & Ones, 1995), meta-analytic structural equation modeling (i.e., MASEM; Bergh et al., 2016; Cheung, 2008, 2013, 2015), and Yu et al.'s (2016) full information MASEM procedure (i.e., FIMASEM). We recommend reading Mackey, McAllister, et al. (2019) and Mackey, Roth, et al. (2019) for details about some of these procedures if you plan to use them in your study. Regardless of the analytical tool you choose, it is important to ensure that the correlations in the meta-analytic correlation matrix are accurate so that the final results are accurate (Roth et al., 2011).

Below, we provide some information about how we conducted relative weight analyses (Johnson, 2000; Johnson & LeBreton, 2004; Tonidandel & LeBreton, 2015) because these analyses likely will be useful and informative regardless of the broader methodological approach implemented. The primary benefit of conducting relative weight analyses is that they identify the amount of variance that the linear versus curvilinear terms predict in the outcome variable. Thus, relative weight analyses are especially useful for our meta-analytic approach due to the concerns about multicollinearity that arise because the curvilinear term is calculated using the linear term.

First, it is important to use a .csv file to create the meta-analytic correlation table described above. Directions for and examples of .csv files can be found at http://relativeimportance.davidson.edu/multipleregression.html. Then, use the .csv file as an input file to conduct relative weight analyses (Tonidandel & LeBreton, 2015) in the R program (R Core Team, 2017; the program is available for free at https://www.r-project.org/) using the relaimpo package (Groemping, 2015). We conduct the analyses directly in the R program, but Scott Tonidandel and James LeBreton also host a website that can be used to run the analyses (https://relativeimportance.davidson.edu/). These analyses produce results that indicate the amount of variance (i.e., R^2) that the linear and curvilinear terms predict independently and together, as well as the percentage of the total explained variance attributable to the linear versus curvilinear terms.

CHALLENGES WITH AND OPPORTUNITIES CREATED BY IMPLEMENTING OUR APPROACH

We encountered several challenges and unexpected opportunities while implementing our approach for conducting curvilinear meta-analyses. We describe the challenges and opportunities below so we can prepare others for the realities of conducting a curvilinear meta-analysis that requires them to contact authors for every primary study to request data.

First, it is important to consider the sheer amount of time and effort that goes into our approach. As mentioned before, we sent personalized e-mails with custom SPSS syntax files to the authors of each and every single primary study we identified for possible inclusion in our meta-analysis. This meant that our preparation for sending e-mails included building the reference section, creating a template e-mail for the data request, creating an SPSS syntax file that was customized for every study, and locating contact information (i.e., e-mail addresses) for as many authors as possible. Even then, we still had to send many follow-up e-mails to people who did not respond to our data requests. In our study, this meant contacting authors of 126 studies with 176 total samples to request data about 352 relationships at least once. We obtained about 59% of the full data requested, so you likely will need to contact many authors to reach appropriate thresholds for power.

Next, it is important to consider the range of responses to e-mails you might receive. There were some interesting responses among the 41% of the data we did not receive. Only about 3% of the authors we contacted indicated that the data were not available. Most of the 41% simply did not respond to our e-mails. A few authors responded with hostile e-mails that accused us of making unreasonable requests. Ironically, many of the authors who sent us data thanked us for making the process so easy for them. Some authors even sent us raw data and told us we could run the analyses ourselves. There were also a few quid pro quo requests made. For example, we were invited to present our research for a different department at the first author's institution. Also, we are pretty sure that it increased our visibility for invitations to review for journals, provide unpublished data for others' meta-analyses, and provide friendly reviews of others' work. Ultimately, it is important to consider that this might be an opportunity to build social networks and increase visibility. However, also consider the time it takes to be responsive to the authors of primary studies who reply to your e-mails because you are probably contacting much of the potential reviewer pool for journals that will review your study for possible publication.

Third, it is important to consider that curvilinear effects often are hard to empirically replicate across contexts. Thus, you will likely have a mix of meaningful curvilinear results and essentially null results even if there are strong theoretical reasons to expect curvilinear relationships. Thus, we recommend examining the final results at one, two, and three standard deviations from the mean. Many scholars use Jeremy Dawson's publicly available plotters (http://www.jeremydawson.co.uk/slopes.htm) to examine the form of curvilinear and/or moderation effects at one standard deviation below and above the mean. Although following this convention helps build cohesive literatures around effects within the middle 68% of distributions, it does not adequately depict curvilinear effects at extreme/outlier values. However, visually examining curvilinear effects at one (i.e., 68% of the distribution), two (i.e., 95% of the distribution), and three (i.e., 99% of the distribution) standard deviations below and above the mean portrays a more complete picture of curvilinear effects. This recommendation is especially important when examining curvilinear effects for variables that are not normally distributed (e.g., destructive leadership) because they may demonstrate strong effects at outlier values. We found that graphs for these three levels often painted different pictures about the magnitude of potential curvilinear effects, which suggests that curvilinear effects may be most pronounced at extremely low and high levels of predictor variables.

Finally, there are interesting challenges in examining curvilinear relationships in general. For example, the correlations for the linear terms can be low when there are meaningful curvilinear relationships. Thus, relationships that get dismissed as too small to be meaningful may actually be hiding curvilinear effects. However, publication bias (Dalton et al., 2012) suggests that these relationships are the most likely to be suppressed, which can make them difficult to find. In addition, there are known issues about the accuracy, reporting, and representativeness of existing results in published studies (Banks et al., 2016; O'Boyle, Banks, & Gonzalez-Mulé, 2017). Overall, the aforementioned issues culminate in our recommendation to carefully evaluate the viability of a meta-analysis that investigates curvilinear effects by examining data from your own studies and studies that are easy to access (e.g., coauthors' and colleagues' datasets) prior to embarking on a comprehensive effort to locate all potential data for inclusion.

FUTURE RESEARCH OPPORTUNITIES

There are numerous actionable opportunities for scholars to leverage our meta-analytic technique while conducting their own research. First, researchers could replicate our use of regression-based analyses (Viswesvaran & Ones, 1995), relative weight analyses (Johnson, 2000; Johnson & LeBreton, 2004; Tonidandel & LeBreton, 2015), and weighted least squares regression (Cohen, Cohen, West, & Aiken, 2003; Lipsey & Wilson, 2001) to examine the ability of curvilinear terms to predict variance in dependent variables above and beyond linear terms. Researchers can even use our procedures to facilitate more advanced meta-analytic techniques, such as metaregression (Gonzalez-Mulé & Aguinis, 2018) and meta-analytic structural equation modeling (Bergh et al., 2016; Cheung, 2008, 2013, 2015; Yu et al., 2016). These approaches can build on contemporary research methods (Cortina, Aguinis, & DeShon, 2017) to improve our understanding of numerous relationships plagued with seemingly conflicting findings across studies.

For example, Mackey, Frieder, Brees, and Martinko (2017) found that there was weak relationship between neuroticism (i.e., a predisposition to be nervous and moody) and abusive supervision ($\rho = 0.12$, $SD_\rho = 0.19$, $k = 12$, $N = 4{,}198$). However, the relatively high standard deviation value (i.e., SD_ρ) indicates that there is heterogeneity in this relationship. Perhaps there are unmeasured curvilinear effects that explain this relationship. Likewise, Wang, DeGhetto, Ellen, and Lamont (2019) found evidence of a weak relationship with a lot of heterogeneity between overall board human capital and Chief Executive Officer (CEO) duality (i.e., combining the CEO and board chair positions into one role; $\rho = 0.01$, $SD_\rho = 0.17$, $k = 43$, $N = 58{,}122$). Perhaps there is a U-shaped curvilinear relationship whereby moderate levels of overall board human capital meaningfully affect CEO duality but low and high levels of board human capital produce similar levels of CEO board duality.

Next, our methods could be leveraged to make advancements for other unconventional meta-analytic approaches. For example, Van Iddekinge, Aguinis, Mackey, and DeOrtentiis (2018) meta-analytically examined the interactive effects of ability and motivation on job performance. Their approach was similar to ours, but they also incorporated techniques that enabled them to meta-analytically examine simple slopes to critically evaluate interaction effects (Aguinis, Gottfredson, & Wright, 2011). Likewise, making concerted attempts to visually depict meta-analytic relationships via traditional plots (Aguinis & Gottfredson, 2010) and Johnson–Neyman plots (Miller, Stromeyer, & Schwieterman, 2013) across the entire range of the predictor variables can provide really useful information that creates nuance to our understanding of relationships. For example, we were able to use Johnson–Neyman plots to demonstrate that there was evidence of a curvilinear effect of destructive leadership on job performance for values of destructive leadership that were below 1.19 and values that were above 2.17 on a five-point scale (Mackey et al., 2019). Spotlights, floodlights, and other visual representations (Spiller, Fitzsimons, Lynch, & McClelland, 2013) of meta-analytic curvilinear data would be informative too.

Although we focused on micro-level management research in our study, our technique can also shed light on the replicability of curvilinear effects in strategic management research (Bergh, Sharp, Aguinis, & Li, 2017). For example, it would be interesting to see whether and the extent to which curvilinear effects differed according to the operationalizations of focal macro-level variables that have been measured in numerous ways. Applying our techniques could provide nuance to our understanding of the boundaries within which curvilinear effects of the features of firms and leadership within firms, as well as how these features are measured, affect important firm outcomes (e.g., financial performance, executive compensation).

CONCLUSION

The purpose of this chapter was to provide a guide for conducting curvilinear meta-analyses by describing the relevant meta-analytic techniques we use in our own research. Overall, our contributions stem from our guide for conducting curvilinear meta-analyses, the useful context for its implementation we provide, and our identification of opportunities for scholars to leverage our technique in future studies to generate novel and nuanced knowledge that can advance their fields. We hope the advanced techniques we described enable scholars and practitioners alike to make more robust inferences about meta-analytic curvilinear effects.

REFERENCES

Aguinis, H., Dalton, D. R., Bosco, F. A., Pierce, C. A., & Dalton, C. M. (2011). Meta-analytic choices and judgment calls: Implications for theory building and testing, obtained effect sizes, and scholarly impact. *Journal of Management*, *37*(1), 5–38.

Aguinis, H., & Gottfredson, R. K. (2010). Best-practice recommendations for estimating interaction effects using moderated multiple regression. *Journal of Organizational Behavior*, *31*(6), 776–786.

Aguinis, H., Gottfredson, R. K., & Joo, H. (2013). Best-practice recommendations for defining, identifying, and handling outliers. *Organizational Research Methods*, *16*(2), 270–301.

Aguinis, H., Gottfredson, R. K., & Wright, T. A. (2011). Best-practice recommendations for estimating interaction effects using meta-analysis. *Journal of Organizational Behavior*, *32*(8), 1033–1043.

Aguinis, H., Pierce, C. A., Bosco, F. A., Dalton, D. R., & Dalton, C. M. (2011). Debunking myths and urban legends about meta-analysis. *Organizational Research Methods*, *14*(2), 306–331.

Aguinis, H., Ramani, R. S., & Alabduljader, N. (2018). What you see is what you get? Enhancing methodological transparency in management research. *The Academy of Management Annals*, *12*(1), 83–110.

Aytug, Z. G., Rothstein, H. R., Zhou, W., & Kern, M. C. (2012). Revealed or concealed? Transparency of procedures, decisions, and judgment calls in meta-analyses. *Organizational Research Methods*, *15*(1), 103–133.

Banks, G. C., O'Boyle, E. H., Pollack, J. M., White, C. D., Batchelor, J. H., Whelpley, C. E. … Adkins, C. L. (2016). Questions about questionable research practices in the field of management. *Journal of Management*, *42*(1), 5–20.

Barnett, M. L., & Salomon, R. M. (2006). Beyond dichotomy: The curvilinear relationship between social responsibility and financial performance. *Strategic Management Journal*, *27*(11), 1101–1122.

Bergh, D. D., Aguinis, H., Heavey, C., Ketchen, D. J., Boyd, B. K., Su, P. … Joo, H. (2016). Using meta-analytic structural equation modeling to advance strategic management research: Guidelines and an empirical illustration via the strategic leadership-performance relationship. *Strategic Management Journal*, *37*(3), 477–497.

Bergh, D. D., Sharp, B. M., Aguinis, H., & Li, M. (2017). Is there a credibility crisis in strategic management research? Evidence on the reproducibility of study findings. *Strategic Organization*, *15*(3), 423–436.

Borenstein, M., Hedges, L. V., Higgins, J. P. T., & Rothstein, H. R. (2010). A basic introduction to fixed-effect and random-effects models for meta-analysis. *Research Synthesis Methods*, *1*(2), 97–111.

Cafri, G., Kromrey, J. D., & Brannick, M. T. (2010). A meta meta-analysis: Empirical review of statistical power, type I error rates, effect sizes, and model selection of meta-analyses published in psychology. *Multivariate Behavioral Research*, *45*(2), 239–270.

Carlson, K. D., & Ji, F. X. (2011). Citing and building on meta-analytic findings: A review and recommendations. *Organizational Research Methods*, *14*(4), 696–717.

Cheung, M. W.-L. (2008). A model for integrating fixed-, random-, and mixed-effects meta-analyses into structural equation modeling. *Psychological Methods*, *13*(3), 182–202.

Cheung, M. W.-L. (2013). Multivariate meta-analysis as structural equation models. *Structural Equation Modeling*, *20*(3), 429–454.

Cheung, M. W.-L. (2015). metaSEM: An R package for meta-analysis using structural equation modeling. *Frontiers in Psychology*, *5*, 1–7.

Cohen, J., Cohen, P., West, S. G., & Aiken, L. S. (2003). *Applied multiple regression/correlation analysis for the behavioral sciences* (3rd ed.). Hillsdale, NJ: Erlbaum.

Combs, J. G., Crook, T. R., & Rauch, A. (2019). Meta-analytic research in management: Contemporary approaches, unresolved controversies, and rising standards. *Journal of Management Studies*, *56*(1), 1–18.

Cortina, J. (1993). Interaction, nonlinearity, and multicollinearity: Implications for multiple regression. *Journal of Management*, *19*(4), 915–922.

Cortina, J. M., Aguinis, H., & DeShon, R. P. (2017). Twilight of dawn or of evening? A century of research methods in the Journal of Applied Psychology. *Journal of Applied Psychology*, *102*(3), 274–290.

Dalton, D. R., Aguinis, H., Dalton, C. M., Bosco, F. A., & Pierce, C. A. (2012). Revisiting the file drawer problem in meta-analysis: An assessment of published and nonpublished correlation matrices. *Personnel Psychology*, *65*(2), 221–249.

Desimone, J. A., Köhler, T., & Schoen, J. L. (2019). If it were only that easy: The use of meta-analytic research by organizational scholars. *Organizational Research Methods*, *22*(4), 867–891.

Dimitruk, P., Schermelleh-Engel, K., Kelava, A., & Moosbrugger, H. (2007). Challenges in nonlinear structural equation modeling. *Methodology*, *3*(3), 100–114.

Disatnik, D., & Sivan, L. (2016). The multicollinearity illusion in moderated regression analysis. *Marketing Letters*, *27*(2), 403–408.

Edwards, J. (2008). Seven deadly myths of testing moderation in organizational research. In C. Lance & R. Vandenberg (Eds.), *Statistical and methodological myths and urban legends: Received doctrine, verity, and fable in the organizational and social sciences* (pp. 145–166). New York, NY: Routledge.

Geyskens, I., Krishnan, R., Steenkamp, J.-B. E. M., & Cunha, P. V. (2009). A review and evaluation of meta-analysis practices in management research. *Journal of Management*, *35*(2), 393–419.

Glass, G. V. (1976). Primary, secondary, and meta-analysis of research. *Educational Researcher*, *5*, 3–8.

Gonzalez-Mulé, E., & Aguinis, H. (2018). Advancing theory by assessing boundary conditions with metaregression: A critical review and best-practice recommendations. *Journal of Management, 44*(6), 2246–2273.

Groemping, U. (2015). Package 'relaimpo'. Retrieved from https://cran.r-project.org/web/packages/relaimpo/relaimpo.pdf

Harris, K. J., Kacmar, K. M., & Witt, L. A. (2005). An examination of the curvilinear relationship between leader-member exchange and intent to turnover. *Journal of Organizational Behavior, 26*(4), 363–378.

Hunter, J. E., & Schmidt, F. L. (2004). *Methods of meta-analysis: Correcting error and bias in research findings* (2nd ed.). Thousand Oaks, CA: Sage Publications.

Hunter, J. E., Schmidt, F. L., & Le, H. (2006). Implications of direct and indirect range restriction for meta-analysis methods and findings. *Journal of Applied Psychology, 91*(3), 594–612.

Jin, L., Madison, K., Kraiczy, N. D., Kellermanns, F. W., Crook, T. R., & Xi, J. (2017). Entrepreneurial team composition characteristics and new venture performance: A meta-analysis. *Entrepreneurship: Theory and Practice, 41*(5), 743–771.

Johnson, J. W. (2000). A heuristic method for estimating the relative weight of predictor variables in multiple regression. *Multivariate Behavioral Research, 35*(1), 1–19.

Johnson, J. W., & LeBreton, J. M. (2004). History and use of relative importance indices in organizational research. *Organizational Research Methods, 7*(3), 238–257.

Kepes, S., Banks, G. C., McDaniel, M., & Whetzel, D. L. (2012). Publication bias in the organizational sciences. *Organizational Research Methods, 15*(4), 624–662.

Kepes, S., McDaniel, M. A., Brannick, M. T., & Banks, G. C. (2013). Meta-analytic reviews in the organizational sciences: Two meta-analytic schools on the way to MARS (the meta-analytic reporting standards). *Journal of Business and Psychology, 28*(2), 123–143.

Lipsey, M. W., & Wilson, D. B. (2001). *Practical meta-analysis.* Thousand Oaks, CA: SAGE Publications.

Mackey, J. D., Ellen, B. P., III, Hochwarter, W. A., & Ferris, G. R. (2013). Subordinate social adaptability and the consequences of abusive supervision perceptions in two samples. *The Leadership Quarterly, 24*(5), 732–746.

Mackey, J. D., Frieder, R. E., Brees, J. R., & Martinko, M. J. (2017). Abusive supervision: A meta-analysis and empirical review. *Journal of Management, 43*(6), 1940–1965.

Mackey, J. D., McAllister, C. P., Ellen, B. P., III, & Carson, J. E. (in press). A meta-analysis of interpersonal and organizational workplace deviance research. *Journal of Management.*

Mackey, J. D., McAllister, C. P., Maher, L. P., & Wang, G. (2019). Leaders and followers behaving badly: A meta-analytic examination of curvilinear relationships between destructive leadership and followers' workplace behaviors. *Personnel Psychology, 72*(1), 3–47.

Mackey, J. D., Roth, P. L., Van Iddekinge, C. H., & McFarland, L. A. (2019). A meta-analysis of gender proportionality effects on job performance. *Group & Organization Management, 44*(3), 578–610.

Miller, J. W., Stromeyer, W. R., & Schwieterman, M. A. (2013). Extensions of the Johnson-Neyman technique to linear models with curvilinear effects: Derivations and analytical tools. *Multivariate Behavioral Research, 48*(2), 267–300.

Ones, D. S., Viswesvaran, C., & Schmidt, F. L. (2017). Realizing the full potential of psychometric meta-analysis for a cumulative science and practice of human resource management. *Human Resource Management Review, 27*(1), 201–215.

O'Boyle, E. H., Banks, G. C., & Gonzalez-Mulé, E. (2017). The Chrysalis effect: How ugly initial results metamorphosize into beautiful articles. *Journal of Management, 43*(2), 376–399.

Palich, L. E., Cardinal, L. B., & Miller, C. C. (2000). Curvilinearity in the diversification–performance linkage: An examination of over three decades of research. *Strategic Management Journal, 21*(2), 155–174.

Pierce, J. R. & Aguinis, H. (2013). The too-much-of-a-good-thing effect in management. *Journal of Management, 39*(2), 313–338.

R Core Team. (2017). *R: A language and environment for statistical computing.* Vienna: R Foundation for Statistical Computing.

Rothstein, H. R., Sutton, A. J., & Borenstein, M. (2005). *Publication bias in meta-analysis: Prevention, assessment and adjustments.* West Sussex: John Wiley & Sons.

Roth, P. L., Switzer, F. S., Van Iddekinge, C. H., & Oh, I.-S. (2011). Toward better meta-analytic matrices: How input values can affect research conclusions in human resource management simulations. *Personnel Psychology*, *64*(4), 899–935.

Shepherd, D. A., & Suddaby, R. (2017). Theory building: A review and integration. *Journal of Management*, *43*(1), 59–86.

Shieh, G. (2011). Clarifying the role of mean centering in multicollinearity of interaction effects. *British Journal of Mathematical and Statistical Psychology*, *64*(3), 462–477.

Spiller, S. A., Fitzsimons, G. J., Lynch, J. G., & McClelland, G. H. (2013). Spotlights, floodlights, and the magic number zero: Simple effects tests in moderated regression. *Journal of Marketing Research*, *50*(2), 277–288.

Tonidandel, S., & LeBreton, J. M. (2015). RWA web: A free, comprehensive web-based, and user-friendly tool for relative weight analyses. *Journal of Business and Psychology*, *30*(2), 207–216.

Van Iddekinge, C. H., Aguinis, H., Mackey, J. D., & DeOrtentiis, P. (2018). A meta-analysis of the interactive, additive, and relative effects of cognitive ability and motivation on performance. *Journal of Management*, *44*(1), 249–279.

Viswesvaran, C., & Ones, D. S. (1995). Theory testing: Combining psychometric meta-analysis and structural equations modeling. *Personnel Psychology*, *48*(4), 865–885.

Wang, G., DeGhetto, K., Ellen, B. P., & Lamont, B. T. (2019). Board antecedents of CEO duality and the moderating role of country-level managerial discretion: A meta-analytic investigation. *Journal of Management Studies*, *56*(1), 172–202.

Wang, G., Harms, P. D., & Mackey, J. D. (2015). Does it take two to tangle? Subordinates' perceptions of and reactions to abusive supervision. *Journal of Business Ethics*, *131*(2), 487–503.

Wang, L., Zhang, Z., McArdle, J. J., & Salthouse, T. A. (2008). Investigating ceiling effects in longitudinal data analysis. *Multivariate Behavioral Research*, *43*(3), 476–496.

Wood, J. A. (2008). Methodology for dealing with duplicate study effects in a meta-analysis. *Organizational Research Methods*, *11*(1), 79–95.

Yu, J., Downes, P. E., Carter, K. M., & O'Boyle, E. H. (2016). The problem of effect size heterogeneity in meta-analytic structural equation modeling. *Journal of Applied Psychology*, *101*(10), 1457–1473.

APPENDIX 1

Template E-mail for Data Requests

Dear Gang, Peter, and Jeremy,

I hope you're all doing well. I'm working on a meta-analysis that examines the effects of negative forms of leadership on followers' workplace behaviors. While reviewing the literature, I came across one of your studies that met the inclusion criteria for my study (please see the reference below):

Wang, G., Harms, P. D., & Mackey, J. D. (2015). Does it take two to tangle? Subordinates' perceptions of and reactions to abusive supervision. *Journal of Business Ethics*, *131*(2), 487–503.

I would like to include your study in my meta-analysis, but I'm asking for your help. Will you please send me a correlation matrix with and descriptive statistics for the variables listed below if the data are still available to you?

I attached an SPSS syntax file to this e-mail that will enable you to quickly obtain descriptive statistics and a correlation matrix for your study if the data are available in SPSS. There is information at the top of the syntax file that will help you prepare the syntax file to run. I request descriptive statistics (i.e., means and standard deviations) for and correlations between the following variables: abusive supervision, abusive supervision squared (i.e., abusive supervision^2), task performance, task performance squared (i.e., task performance^2), interpersonal deviance, and interpersonal deviance squared (i.e., interpersonal deviance ^{2}).

Please let me know if you have any questions. If possible, please send the requested information within the next month. I will appreciate any help you can offer.

Thank you for considering my request!

Jeremy Mackey

Assistant Professor of Management

Harbert College of Business

Auburn University

jmackey@auburn.edu

APPENDIX 2

Template Syntax File for Data Requests.

Wang, Harms, and Mackey (2015):

```
COMPUTE AbusiveSupervision = XXXXX.

EXECUTE.

COMPUTE InterpersonalDeviance = XXXXX.

EXECUTE.

COMPUTE TaskPerformance = XXXXX.

EXECUTE.

COMPUTE AbusiveSupervisionSquared = AbusiveSupervision*AbusiveSupervision.

EXECUTE.

COMPUTE InterpersonalDevianceSquared = InterpersonalDeviance*InterpersonalDeviance.

EXECUTE.

COMPUTE TaskPerformanceSquared = TaskPerformance*TaskPerformance.

EXECUTE.

CORRELATIONS.

/VARIABLES = AbusiveSupervision AbusiveSupervisionSquared InterpersonalDeviance InterpersonalDevianceSquared TaskPerformance TaskPerformanceSquared.

/PRINT = TWOTAIL NOSIG.

/STATISTICS DESCRIPTIVES.

/MISSING = LISTWISE.
```

PROCESS RESEARCH METHODS: A CONVERSATION AMONG LEADING SCHOLARS*

Raghu Garud, Paula Jarzabkowski, Ann Langley, Haridimos Tsoukas, Andrew Van de Ven and Jane Lê

Keywords: Process; process research; process research methods; organization theory; dialogue; research philosophy

Process studies have become an integral part of organization and management research. Theoretical contributions to understanding, exploring, and studying processes (e.g., Garud, Berends, & Tuertscher, 2018; Jarzabkowski, Lê, & Balogun, 2019; Langley, 1999, 2007; Tsoukas & Chia, 2002; Van de Ven & Poole, 2005), a cumulative number of special issues (e.g., *Academy of Management Journal*, 2013; *Organization*, 2002; *Strategic Management Journal*, 1992, 2018), and annual process-centered PDWs, conferences, and symposia (e.g., the *Process PDW* at *AOM*; *PROS*) mean that process studies have gained considerable momentum in recent years. Yet, despite significant theoretical and methodological progress, ambiguity continues to surround the term *process* and the practice of *process research* (Sandberg, Loacker, & Alvesson, 2015). Built from a panel discussion, this chapter features reflections by five key process scholars in order to draw out different understandings of process and the implications of these understandings for ways in which we conduct process research. We structure the chapter along the format of the panel, starting with a brief reflection by each

*This chapter contribution has been crafted from a transcript of a Panel Symposium (#1028, "Theorizing Processes and Process Theories in Organization and Management Research") that took place at the Academy Management Conference in Chicago, Illinois, in August 2018. The original symposium was organized and facilitated by Eric Knight (Macquarie University), Jane Lê (WHU – Otto Beisheim School of Management), Sarah Stanske (Leuphana University), and Matthias Wenzel (Leuphana University).

Advancing Methodological Thought and Practice
Research Methodology in Strategy and Management, Volume 12, 117–132

ISSN: 1479-8387/doi:10.1108/S1479-838720200000012019

process scholar, before "going into conversation" prompted by questions in order to allow the crossing of perspectives.

SECTION 1: REFLECTIONS

In order to guide the reflections, each scholar was given three initial questions as prompts: (1) what is your perspective on process, (2) what are the methodological implications of such a view, and (3) what are the most significant opportunities for process research studies right now?

Raghu Garud, Pennsylvania State University

My own journey into process research started as a PhD student in 1984. Because of my prior background, I thought of process from an engineer perspective – an input, a process, and an output. Yet, this view of process obscured the social issues involved, which, to me, remained a black box. Moreover, in pursuing this notion of process, I still clung to variance theorization wherein variability in outcomes could be attributed to variability in inputs, with processes serving as mediators.

This all changed one day when I met Andy Van de Ven, who told me, "Go beyond variance theorization and start thinking about process theorizing." I asked, "What is process theorizing?" He said, "You must ask questions about *how* things happen." And, as a consequence, I started thinking about and studying events and about how they unfolded over time (i.e., patterns). This was a big shift from "process as black box" to "process as a pattern of events," a move that essentially endogenizes the contexts within which processes unfold.

Then I met other scholars – Hari Tsoukas, Ann Langley, Paula Jarzabkowski, for instance – and found that their interests around temporality matched my own. They, as was I, were thinking about how people understood time. Once you start doing that, you go beyond process as a pattern of events to process as how people experience phenomena. An appreciation of how people experience phenomena endogenizes the notion of time itself.

I am sharing these observations with you to note how far we have come with the notion of process – from the notion of process from engineering, to the notion of process from an appreciation of events, to a notion of process as phenomena being experienced. Each notion of process implies a different onto-epistemological package. To the extent that one views process as experienced using a distributed ontology, for instance, then, the onto-epistemological package is different from the one implicated in other notions of process. If research on process as experienced is reviewed and evaluated by scholars who take a different onto-epistemological position, then, there is a real risk of the work being selected out. Consequently, if you are not careful, then a lot of hard work could be rejected for the wrong reasons.

Paula Jarzabkowski, City University London

I first ought to clarify how I see process. There are two key principles in that for me. As a practice scholar, the first thing that I'm interested in is the *doing* or what I would call the practice. In other words, what people are doing at any moment in time. And doing can be really, quite mundane. In the research that we've done, we've seen how doing even little things – like people putting on ties and knocking on doors – can be critical. Of course, these things are not terribly interesting in isolation. We don't want a theory of putting on ties. We want to understand more than that.

For me, the notion of becoming is also very important. It is where I see the unfolding of doing over time and the *interconnection* of doing over time. From my perspective, it is this continuous interplay between what we're doing at any moment and what that doing is allowing to become, or the unfolding flow of experience that arises from that doing. I do think that we have come a very long way, particularly in terms of actually having that language and being able to use it. "Doing" and "becoming" is an acceptable language.

That certainly wasn't always the case. In the past, it was still very much about phase models and explaining shifts between phase one and phase two. And, actually, methodologically speaking there's still a lot of value in using phases as analytic devices in order to try and understand what this becoming is that you're trying to talk about. Continuous flow and experience are very hard to talk about. One of the challenges that I repeatedly encounter working with a worldview in which things are always in transition, always unfolding, is that I still need some texture or tension in order to show what is unfolding. It's hard to define a river if there are no river banks and no ways to see in which direction the river flows.

An important question for me is where to go next. I personally would like to see process theory realize itself even more strongly. It's not a theory of things that just happen in organizations. Process is the world we live in and this view has the potential to allow us to address a lot of big, social issues – things like poverty, climate change, risk, inequality, and so on – these really significant interorganizational, intertemporal processual phenomena. These simply cannot be studied any other way than strong process because they are complex and they move between people and things and across time. But rather than seeing it as a frustration, I think it's a real opportunity. And I'm very excited about the opportunity to try to do some of that. I do think that some of the process methodologies we have now enable us to at least explain some of those things. And, if we can explain them, then there is hope that we can address them.

Ann Langley, HEC Montreal

I would like to start off by reflecting on why we are even bothering with process. It is very important to say that this is not just an intellectual exercise that we are engaging in for the fun of engaging in an intellectual exercise. We are looking at process because process is important, because temporality is important, because time is important – that's the one thing we can't escape. A lot of our research is looking at how things affect outcomes, but a process perspective actually enables us to see that this notion of outcome, it is a false concept in a certain sense. Since time doesn't stop, there is no ultimate outcome. That means we need to think about how things continue to go on after that moment of "outcomes."

In a recent book chapter with a student of mine, Fernando Fachin (Fachin & Langley, 2018), we suggest four ways of thinking about process, rather than using the standard distinction between strong and weak process. The first approach we called *process as evolution*. This corresponds to the idea of process as implying change in things. Things exist and what we are looking at is their evolution over time. There are some ideas about how to study this in a paper that I wrote two decades ago (Langley, 1999). This type of process research essentially builds on a substantive ontology (where the world is assumed to be made up of substances) and we pretty much know how to study it methodologically. So, while it is extremely important in order to understand how things evolve over time, this is not where the frontier lies in process thinking. There are more challenging and interesting ways of thinking about process, which are worth considering.

The second approach we called *process as narrative*. This builds on an experiential ontology, where we are looking at process as conceived by people. So, what we are studying here is the stories people tell about process as they experienced it. That means that the kind of data that we are going to be using is likely to be different. We would likely be looking at interviews because we want to capture how people understand process. This approach is also quite well-established and so I also do not think of this as where the challenges really lie.

The next approach, *process as activity*, is something that Paula's work really brings to life: the idea that the world is made up of processes of doing and becoming (a process ontology). Here we need to use different kinds of data and I think we are learning how to do this right now. This involves getting much closer to the actual interactions occurring among people and what those interactions are accomplishing. Not the outcomes, in the variable sense, but what the activity that we engage in is achieving. This is a different way of thinking about process and outcomes. I think that we are beginning to understand how to do this kind of research, and it has begun to penetrate the top journals. There is a great deal of potential here. For instance, one thing that one can also do with this approach is to understand the processes involved in staying in the same place: to understand how certain patterns of activity are continuously reproduced. Those kinds of processes are also really interesting.

The fourth and final approach to process Fernando Fachin and I discussed in our chapter is *process as withness*. This is difficult to engage with as it involves taking process philosophy to its logical conclusion. If you think about it carefully, even when we engage in research that focuses on *process as activity*, we assume that the researcher is outside this process. However, if we think more radically, and accept that the world is continually in process, then we too are in process along with it. This approach recognizes and moves closer to that notion by conceiving of us researchers as moving with those that we are studying. Doing this type of research becomes quite challenging because, on the one hand, we now have to think about being with others as we study them, and also to think about studying forward rather than backwards. Yet, most of our research is still about what happened in the past. Finding an approach that allows you be in movement with the participants, doing research forward, rather than backward, that is the real challenge.

Haridimos Tsoukas, University of Cyprus & University of Warwick

In the Introduction to the *Sage Handbook of Process Organization Studies*, Ann and I distinguished two waves of process research. The first wave comprises work up to the 1990s, followed by the second wave of process research. These would roughly correlate with a "weak process" approach and a "strong process" approach. I caution myself here as weak and strong, for me and Ann, are purely descriptive not evaluative terms. They incorporate different ways of viewing change. In the first wave, we see change mostly modeled on motion, whereas in the second wave, we see the effort to understand change in more complex terms – as immanent.

I align myself with a strong process view or at least a stronger view of process. Thus, I'd like to highlight three key features, which I think are important if you take a strong process view. First, an interest in experience. Agency is very important in process and the experience of the agent is very important. And, in fact, experience changes the agent along the way. All our colleagues here have done empirical research to show this. Hence, capturing experience is very important empirically.

Secondly, heterogeneity matters. When I talk about heterogeneity, I mean that no two experiences can ever be the same. Think about how many times you have sat in endless academic meetings and felt frustrated. Even though you felt frustrated many times, it's never the same experience of frustration. That is because anytime we experience something, that modifies us, changes us, however imperceptibly. The question is thus how we can capture that heterogeneity that is immanent in the experiential understanding of life.

Thirdly, temporality is critical. Experiences evolve over time, so we need to find ways to capture the temporal constitution of organizational life. We need to distinguish class a distinction between chronological time and experiential time, and I think we need to elaborate those distinctions in order to capture the temporality of life. The thing that sets apart process research is that we get away from the discreteness of life, not seeing life as episodic, as a simple sum, or as succession of discrete events and objects. Rather, we are trying to understand their interpenetration; the way they comingle. And I think this is a very relational image that is difficult to carry out empirically.

Some people might argue and might suggest that "discreteness in life is all that surrounds us." Pragmatically speaking, they would be right. When I drive, I need to be very pragmatic in order to avoid the truck, which may be coming onto me. And if I don't see life in those discrete terms, I may soon disappear. However, just because we tend to view life in discrete terms, and this mostly serves us well, it does not mean that this is all there is in life. I see process research as trying to uncover those portions of life, if you like, which are hidden from ordinary view.

My final point is a plea. I want to make a plea for process and practice streams to come together. I want to see process and practice intersect. The logic of process is becoming. It is never to see life as complete. It is an ongoing project. The logic of practice is the logic of entwinement. People are embedded into broader, collective practices and ways of understanding the world, which are not necessarily aware of, but nonetheless they are material because people necessarily draw on those understandings to act in the world. If we find ways of joining forces, bringing together theories and methods from process and practice, then this is a promising way forward.

Andrew Van de Ven, University of Minnesota

Scott Poole and I are currently editing a second edition of the Oxford Handbook of Organizational Change and Innovation. It focuses on the many new concepts and findings that have emerged since the first edition in 2004. Like the first edition, the handbook focuses on processes of change, or the sequence of events in which organizational characteristics and activities unfold over time and the factors that influence these processes. Across the diverse 31 chapters in the second edition of this handbook, two basic questions consistently present themselves. First, what is the nature of change and process? New processes of organization change are emerging in many forms, including planned and unplanned, episodic and continuous, incremental and radical, alternative generating mechanisms or motors, and stability in changes. Second, what are the key concepts in theories of change and innovation? They include human agency, time conceptions, causality, and levels of analysis. In addition, we add voices not heard in the first edition about affect and emotion, power and influence, paradox and conflict, and critical and political perspectives on organization change, creativity, and innovation. As these new concepts and findings suggest, now is a very exciting time to be a process research scholar.

SECTION 2: IN CONVERSATION

This section reflects conversation between the scholars, prompted by questions from the organizers and the audience, in order to facilitate dialogue across perspectives and draw out important nuances inherent in the various approaches.

On the Number One Issue in Process Research

Hari: I can only see it from my vantage point. I was recently on a session on the age of the Anthropocene. It was an eye-opener and I think that it is the number one issue for us, in terms of what's happening to the planet, what's happening to our world. So, we cannot think in reductionist forms, our thinking has to be holistic, it has to be systemic. The other key issue is regeneration, specifically how people can regenerate. In terms of process, in terms of becoming, these are crucial issues.

Ann: I agree with what you are saying, that if we really want to understand a world where everything is connected to everything else, then part of the challenge is being able to study those kinds of things. The focus is not simply going to be on processes located in small places, but often it needs to be on processes that intersect at multiple levels and do so globally. I'd like to pass the microphone to Paula to tell us a little bit about how she is doing this kind of work.

Paula: When you do that, the thing that you're going to have to accept is that process theory is a theory of contradiction – it's dialogic, it's dialectic, it's paradoxical. That means you need to follow the contradictions that emerge because of these interconnections. These complex problems arise from interconnections between actors who don't even know that they are connected. We have the tools as process scholars to follow where contradictions bubble up, to look at problems, paradoxes, and how they shift between organizational actors. If we do that well, then perhaps we can help those actors to center themselves, people who think "it's not my problem" to help those people connect the dots. Process is a lens that can help do that.

Ann: At the same time, we need to focus on the small and how things get accomplished in the moment. So combining those two is one of the challenges. The *process as activity* perspective that I introduced earlier requires us to look at the process in the moment, but by doing that we may lose the context – and we need both.

Paula: Absolutely, you can't study a global problem – where is "global"? It is instantiated in the local. That means that you have to look at local things and how they connect to other local things, until it is no longer "local."

Hari: For me, the main challenge is to capture life from within. This is easy to say, difficult to do. You can do the process in a reconstructed way, i.e., something happened, a merger, a change, whatever, and you look back. That is relatively easy and we have been doing it for years. But how do you report on life as it unfolds? How do you capture things in the moment as they emerge? Empirically it is difficult. There are ways of trying to do this. This is difficult but really quite important research. It will enable us to capture the heterogeneity of experience by zooming into the moment, into the way things emerge and flow, as well as time. Understanding time and temporality is profoundly important. And getting more sophisticated ways of doing this is critical to push the field forward.

Andy: I would like to add that process is all about know-how, not know-what. Variance models focus on causal factors that explain the "know-what." Many managers and organizational participants are asked to apply "know-how." They are told to "implement and change this or that," but they often don't know how to do so. They need knowledge of the process by which things unfold, the steps, the sequence, and the stories, of how to do tasks. We as scholars have tended to view process in a somewhat pejorative way as representing a proletariat form of know how that practitioners should be able to figure out. I don't mind addressing proletariat tasks that are fundamentally needed and help organizations function.

On Pushing Process Studies Forward Empirically

Andy: Process studies focus on observing processes over time in terms of the occurrence of events, activities, or changes over time. Some of the most challenging tasks in empirically observing a process is defining the events or activities and the decision rules used to code them. We need to identify the concepts that we're looking at. We can't see everything. So when we're taking a qualitative or a quantitative approach, there is an important need to focus on the issues and to be clear about the concepts being observed over time. We then can conceptualize these as events or activities or incidents that people take, and we track the activities over time. Process is all about what happens over time, not across cases.

Hari: I think my colleagues here have done wonderful research, so they can offer more insightful advice, but what I can say is that language matters a lot. Trying to capture the nuance of meaning, how new meanings emerge is vital. These are tricky points because new meanings imply new types of action, new possibilities, and that would then mean that people aren't afraid to act on those possibilities.

Raghu: As a scholar trying to understand the emergence of cochlear implants, I had to take some innovative approaches to collect field data. For instance, I used conferences as field configuring events. In a paper on this topic (Garud, 2008), I wanted to bring the reader into the conference along with me. So, what I did was to include photographs, texts, scribblings of people, and other artifacts from the conferences I visited into the paper to bring these events to life for the reader. Conferences are confusing, ambiguous, emergent phenomena. But, the problem is that we can't write it in that manner. When we write the account and send it to journals, it becomes very linear, neat, and clean. Such accounting does not capture the nonlinear ambiguous experiences of the participants at the conferences nor is it true to the process of doing process research. So, when we think about what it means to write about process, I would say that we need additional formats. For example, what Andy did with the *Academy of Management Discoveries* provides scholars the opportunity to bring in videos and other new forms of data, and to play around with different ways of presenting research.

Ann: In relation to the four perspectives I introduced, I think the challenge is really with the *process as withness* perspective. The question is how to address that. I feel that the people who are doing action research are moving into this space because they are looking forward together with the people they are working with and trying to assist them during that process. Of course, the challenge is that the legitimacy of this kind of work in our journals is not fully established and that acts as a counterbalance. I also think that there are marvelous opportunities with the *process as activity* perspective, given the availability of data and the new ways of collecting data that allow us to actually capture activities in the moment: interactions in the moment, recorded e-mail conversations, social media, and all sorts of interactive traces that you can now capture that we could not do before. I believe this provides great opportunities to access and understand how things unfold in ways that perhaps was not possible before. We should really be looking at these opportunities and seeing how we can understand processes differently with these materials.

Paula: I'm going to start off by saying that I *love* data. I love the puzzle and trying to understand how can we find out more. For me, the great thing about process research is that there's no bounds in research. You find a puzzle and you try to follow it. Process research is a wonderful method for that. Let me give you an example from my own field work. One of the things we're looking at is a global market. The thing is, we never actually set out to study a global market. What happened is that we started by comparing local markets and realized that it isn't possible to compare two markets, precisely because markets are global manifestations. And if that is the case, then that has implications. For us that meant having to find ways to follow the "local" in multiple situations and learning how to do multisite, multimember ethnography because it was all connected.

Our question was, what would it look like empirically? It would look like those things we talked about. We have to have a flat ontology. That is because we're not saying that the local is the global, but at the same time recognize that you can't possibly understand why something is happening in a particular way in London and another way in Bermuda if you don't understand how it happens as a capital market that flows relationally between all of those contexts. A flat ontology means that I'm not comparing London with Bermuda – this is not a comparative case study. Instead, we're trying to see how things connect in the real world. We have such amazing opportunities in process research and we are allowed to be creative in our methods. For instance, right now we're trying to do another study of this kind, but actually making it even bigger and more complicated, with more types of actors who span contexts. This is really exciting. We're playing around and are trying to find out how we're going to do a study like that. The techniques of ethnography and other methods are there, but it's up to us to turn them into things that let us access these problems we want to access.

Where do you normally find the best process accounts? If you read a good novel, you cannot help but get distracted by the subtlety of description. Why? Because there's a lot of richness and interconnectivity, and there's a lack of analytical categories. One of our big professional handicaps is our language – how can we learn to do better? To be more artful and creative in our writing? We need to ask ourselves when it will help our findings, when it will be able to capture the flow? What makes the most sophisticated analytical categories more sophisticated? We need to get better at grappling with this contradiction. The fact is that this contradiction will not go away. If you try to be analytical, you're bound to lose some richness, and the more richness you retain, the less analytical you'll be. The trick is to find the happy medium.

I also think we need to be pragmatic, particularly in the way we advise early career scholars. We want to do great process research and also get published. So my advice to anyone doing a PhD at this moment would be to say: "I never did a global study. I did my study of A, and then my study of B, and my study of C, and they all joined up." Do an amazing study of something you can get your hands on, and just collect as much data as you can, because you may not know what's important in that moment. The best advice I ever got when I was doing my PhD was from Ann (Langley). I sent her an e-mail asking for advice on a paper and she was very nice because she responded to my e-mail. She told me the paper was good and even gave me detailed feedback to help me improve it. I love that. That gave me the confidence to submit the paper. It eventually became the first publication from my thesis. And there are always good things like this that happen. You may not always get exactly what you want, but you'll always get something, even if it isn't what you originally set out to find. Collect all the data you can and look after it very well. Know where it is, record it well, and have access to it. You might not use it for years, because you're busy collecting other datasets, but keep it. You may surprise yourself by returning to it many years down the line, looking across all the data you've collected and think: "Oh, I finally understand." That would be my real advice: Collect each piece of data discretely, love that data, and always come back to it.

Ann: I absolutely agree with everything that Paula has said. Never think of a project as the last thing you are going to do. Your project is a part of a long career and there will be opportunities to accumulate material if you can think programmatically about your work. At the same time, for your thesis, you need to think discretely and have a small, manageable project that you can complete. I talked about looking at *process as activity* already and I think there are wonderful opportunities for doing that type of work without needing huge amounts of data and lengthy time periods. You can look at activities in "doable" year-long studies that still allow you to examine important processes. The other pragmatic bit of advice I would give you is that if you do want to look at something over a long period of time, then start today. Don't wait! That way, by the time you are defending your thesis, you will probably have 2 or 3 years of data. There are examples of students doing that. So you need to decide early what you want to do and begin do it.

Raghu: Ludwik Fleck made an important point, which resonates with me and might be helpful. He talked about *thought styles and thought collectives*. Thought styles are ways of thinking informed by thought collectives or epistemic communities. Thought collectives emerge based on a tradition of research around the beliefs that scientists hold, the judgments they bring to bear, and the instruments they use. Applying Fleck's insight to process research, it is important to understand the kind of process research you want to undertake, and to which thought collective you want to belong. In line with this, I suggest you submit your papers to journals/editors that are sympathetic to your thought style. So that's the first piece of advice I would say in terms of trying to get yourself published.

The second set of concepts that I find useful is the "fallacy of processification" and the "fallacy of reification" (Thompson, 2011). If your methods capture snapshots, and you are trying to create a process theory out of that, then, that is problematic. Whichever journal you send your paper to is going to say that your epistemology and ontology are misaligned. On the flip side, if you've got longitudinal data and you are trying to make it into a snapshot or a causal model, then you will also run into trouble. My suggestion is to make sure that your epistemology and ontology are aligned.

Andy: Unlike 20 years ago, process research is now considered legitimate science in most management journals. I don't think management journal editors will be surprised to see process research papers and, if they are relatively open-minded, to appreciate studying process questions as much as the mainstream cause–effect studies with variance models.

To echo what Raghu said, please do consider *Academy of Management Discoveries*. It is a wonderful outlet for thinking processually and writing in a processual way. It begins with a simple description of what you saw unfolding over time. Then, you introduce your puzzle. That is, the thing you or the field didn't know. I have observed this funny trend to abductively come up with a hunch. Hari [Tsoukas] pointed out what a process theory is in 1989. From a realists' point of view, he suggested that a process theory is an explanation for an observed set of events, activities, and incidents, with a search for generative mechanisms that explain why it happened at that particular time and place. That is a process theory. So the end of your paper focuses on your explanation of why this observed set of events occurred in a particular context.

Also remember Anne Huff: Writing is a conversation. So, interact with and talk with your intended audience. You're not writing to a nameless, faceless journal. You're writing to a particular set of scholars in this community, who are not only scientists, but who are also teachers, who are also practitioners, who are engaged in a wide variety of different roles in life. Search out and talk with a few of them before your write your paper. It provides you an understanding of what it is that's important, what assumptions they have, and what concerns they are dealing with.

On the Role of "Relations" in Process Research

Raghu: Gender studies and science and technology studies deal with relationality. One example is Karen Barad's focus on intraactions. It is through the entanglement between things and people that our identities shift and the functionalities of materiality emerges. The very notion of *agencement* (i.e., agency in the arrangements) speaks directly to your point. To all of this, I will add not just ontology but also axiology, by which I refer to the values implicit in our theorization. I think that we are not giving enough attention to axiology. The fact is that different people around the world all have different matters of concern. When we say something like "sustainability," it means something very different to people in India as compared to people in America. We need to understand that.

Hari: I agree that axiology is very important. It's the study of values. This goes back to my point about practices. Only once you understand that the agents embedded in practices are being driven by certain values, then you can begin to understand what it is that they do. I also don't believe that experience is reductive. Far from it. To try to capture the experience of a human being is extremely important. It is not just a purely cognitive or emotional matter. You have to understand the way agents are constituted to be receptive of certain kinds of experiences. That means that the broader community and the broader context will be brought in. When historians tried to understand the French Revolution, for example, one of the things they sought to understand was how the notion of public opinion emerged in eighteenth century France and how new texts, books, and pamphlets published at the time modified people's experiences. These are very large-scale questions. But I think you can bring them down to a kind of lower scale, where you try to see the way people experience things, in particular conjunctures of time, and then try to map it out and to capture it.

Paula: I do agree with that very much. In microresearch, the issue is always whether and how to reduce things. Yet, the very act of being human can never be anything but at the nexus of all this – it's the social, and that means the large is writ small in anything you can do. You can't be human and not interact, you can't even have a thought that is not automatically relationally connected to the language we use. The possibility of thinking is structured through connectedness, so I diverge with Hari on this point.

On the Notion of Flat and Tall Ontology in Process Research

Andy: Flat means that you see everything at the same level of analysis, the same unit of analysis – individual rather than group, organization or larger society. Tall means that you're cutting across those levels. Indeed, process studies really need to go across these levels. That is because some of the events and activities are determined and driven by or conducted by institutions. For example, the US federal government has laws that tell you to do things and if you don't do these things, you will go to jail. In this example, I would have to connect the federal government and its institutions of enforcement practices with individual behavior. Actors in various events can represent individuals, groups, organizations, industries, and national governments. Empirically, we deal with this diversity by calling them the actors in an event.

Raghu: If you're not careful, we begin thinking of what happens at "the top-level" serving as context when we take a tall ontology. My position is that we don't operate within contexts; instead, we contextualize. A flat ontology is one in which the government is an actor in its own right. Thus, I believe, the flatter the ontology, the better. I do come from a science and technology studies perspective where scholars try and flatten everything. That's important because, if you're not careful, with a tall ontology, everything becomes a context – something that occurs "out there" – and that's not a fun place to be.

I believe we need a new way of thinking that allows us to see the whole in its part. Gareth Morgan wrote about that in his book "*Images of Organization*." He stressed a holographic approach, which is the capability of each constituent part to contain the whole. Yet, in large, these ways of thinking and associated languages are not available to us. And, so, we sit here on this stage asking about the units of analysis and talking about which variables are changing. But that takes us more and more toward a reductionist way of thinking. That is not going to take us to a more holistic ways of thinking.

Ann: I find a flat ontology very attractive intellectually. But we also need to be pragmatic. At some point, you are going to write a paper and are going to have to make cuts somewhere. That means that you need to pragmatically identify what you are going to focus on. And sometimes identifying levels is useful. It does approximate part of what we are trying to understand. A flatter ontology would give you a different process explanation, but this is also very difficult to do. I would thus invite everyone to try all kinds of ways, of dividing up the world to better understand it. I don't think there is a problem with doing that. But there is something very attractive about someone who really manages to keep the ontology flat. Because there is something special about the sum of the stories that you can see and how small things are actually big things. It is really quite fascinating how you can see that the little conversation in the moment is producing something which is actually quite significant. That's really an attractive thing to do but it is not for everyone, and that is okay. We can still do process research by chopping things differently.

Paula: One of the things that I'm always looking for is theoretical ways of showing how everything is connected to everything else. But, as Ann said, it's not helpful to try and write a paper about everything being connected to everything. That means you also need the social theories that have a flatter ontology that allow you to look at something in a particular moment of time via particular people and show that this is the nexus of all doing. You do need to be able to do that carving up. People like Ted Schatzki were very influential for me when I was looking for theories. After all, we're looking for ways how we want to explain the world.

Raghu: Maybe it will also help to give an example of flat ontology. Let's compare the view of narratives-as-performatives to narratives-as-representations. In the performative view, narratives are intertemporal and intertextual. This is a flat ontology. Paul Naipaul wrote "*India: A Million Mutinies Now*." He went to Mumbai and studied social movements by asking everyone to share stories of their grandparents and to reflect on what their grandchildren's lives might look like. As people recounted their stories, Naipaul could see their life histories and aspirations. You start to see the relationality and temporality through these stories, as they are "holons," i.e., parts that contain the whole.

On Hesitation around the "Process-As-Withness" Approach

Ann: It is very philosophically attractive, but also very hard to do. That is my reservation – It would be hard to do well. To really take seriously the idea that the world is moving forward is difficult when we write papers and, once written, papers are written and thus always in the past. That means that there is a kind of illusion about always being able to move forward. A withness perspective would thus actually mean not writing papers. Instead, you would work with people thinking forward all the time. If you really want to take that approach to research, then action research does some of that, but even then it is still always in the past by the time it gets written up. This approach thus begins to question the potentialities of research. If you take the philosophical perspective, then it represents its own logical conclusion. There is a lot of potential, but we are not quite there yet.

Paula: Yeah, I was really intrigued by this notion of withness. I think it is there when you're doing your research, in this very process of foregrounding and backgrounding. So, I would like to talk about how, if I wanted to do that kind of research empirically, I would do that. I start with a simple observation. We're *always* "with" when we're in the field. That means dealing with emotional content. For instance, when I'm working with people, we send each other research e-mails, trying to make sense of it. But if you're on your own, you might not be able to do that and you certainly won't even understand it all. In both scenarios, you are dealing with your "withness" – being there in the field and reflecting on how you feel about it. It's not always about you trying to influence the labyrinth, it's just about you being there and with them as it happens. But I tend to write that out. I don't think I'm writing a theory about me and how I felt, or me and what I did. That's just because I've always been interested in something else as a phenomenon, and pragmatically I make the choice to let the other stuff go. Others might bring that in, and they might have more "with" accounts. As a researcher that has spent a lot of time in the field, it's still very alive for me. Even months and years after I have lived it, after I have written it – when I reread it, I'm back there. For me, "with"ing is still happening when the period of collection is over. It's still happening in my office when I'm doing my analysis.

Ann: I would argue that every time you go back to your data, you would be with it in a different way. You would be of a different frame of mind because you have changed since the last time you moved through the data. It could have infinite regress. That is why I have a little reservation about it. But I also believe that we can approach it.

Raghu: Once a paper is finished, it's no longer ours. I'm personally quite persuaded by scholars like Umberto Eco who highlighted the role of the reader. We need to understand that our papers may be read by many other people down the generations. People come to me and tell me things about my papers which I never thought about. Indeed, I cannot and do not want to write my papers in such a way that it removes interpretative flexibility. Years down the track, someone might read my work and say, "Your work made me think in this manner given my context." This is an important part of the scholarly process.

Hari: Withness thinking is a difficult notion. It gets often misunderstood. When John Shotter, who introduced the notion, refers to withness, he does not refer to it in order to theorize *about* people, but rather to affect change in people by "being with." With this thinking, it's all about being very sensitive in the moment, if you want to effect change in something.

The question is how we can make use of this notion in our field? One way is actually through empirical research. For example, finding ways of trying to capture the way your agents change and modify their use. You are part of that process of change as a researcher because you're instrumental in the whole process. However, when I write about the phenomenon as a scholar, I'm not writing about myself. I myself am coconstitutive of the phenomenon I investigate, but nonetheless the paper or book is not about how I feel. In that sense, we've got to make some pragmatic distinctions.

We also need to acknowledge the institutional game that we play – and I mean that in a nonfrivolous way. In our scholarly game, we are not so much focusing on ourselves, like a novelist might be focusing on herself in the way she will be writing about, say, the impact of the Chernobyl nuclear accident. In the kind of scholarly game we play, we are focusing on the world outside of us, while simultaneously being mindful of our effect on the world. The phenomenon we experience and study is not independently out there, but we're coconstituting it. That means we have to be mindful of our role in it. At the same time, it's not about us. In that sense Raghu is right – the battle is never finished. When I read your paper and take it into my life, and I'm influenced by it or the community is, then the paper extends its life. But from the point of view of the paper itself and the author, the paper is done.

One of the things that we often confuse in process research is what the phenomenon is. What are you focusing on? There are certain things you don't take for granted, things you need to explain because you're puzzled by them. But you cannot begin to interrogate everything. Certain assumptions are made, certain things are taken for granted, in order to focus on something else.

On the Value of Doing Process Research

Hari: Our quest is to understand how things unfold, how things are done. If you look at our history over the past 40 years in social science, you'll see that we have largely focused on behavior. In a sense, the practice perspective is grounded in that point of view. But then, in about the 1970s, the cognitive turn came along. And all of a sudden it was no longer about what you do. It became about what you think. That explained how and why things happen. "How I think" is interpreted in many different ways – cognitive framing, scripts, narratives, and stories. That made sense. It provided a way of understanding what we do. However, that is not adequate for addressing a lot of other issues. Especially when we have emotions like love, hate, or revenge. All approaches generate mechanisms for why people do what they do. Yet, even that doesn't solve a lot of other issues about belonging, about being, about withness. Maybe the future goes more toward a faith, a vulnerability and a goodwill of others. In a way, our perspective is like a religion in that it provides an explanation for how and why I do what I do. It allows us to begin to think. In the future, I believe we need to appreciate the important role of religiosity in management and organizational behavior processes. I'll never forget Reverend Leon Sullivan, the architect of the principles against apartheid in South Africa and a pastor of the Zion Baptist Church in northern Philadelphia, observing that people come to church on Sundays in order to get their pitchers filled. Their pitchers are a quest, a desire – something has to fulfill their need for being. For me, that quest is process research.

REFERENCES

Fachin, F. F., & Langley, A. (2018). Researching organizational concepts processually: The case of identity. In C. Cassell, A. L. Cunliffe, & G. Grandy (Eds.), *The SAGE handbook of qualitative business and management research methods: History and traditions* (pp. 308–327). Los Angeles, CA: SAGE Publications.

Garud, R. (2008). Conferences as venues for the configuration of emerging fields: The case of cochlear implants. *Journal of Management Studies*, *45*(6), 1061–1088.

Garud, R., Berends, H., & Tuertscher, P. (2018). Qualitative approaches for studying innovation as process. In R. Mir & S. Jain (Eds.), *Routledge companion to qualitative research in organization studies* (pp. 226–247). New York, NY: Routledge.

Jarzabkowski, P., Lê, J. K., & Balogun, J. (2019). The social practice of co-evolving strategy and structure during mandated radical change. *Academy of Management Journal*, *62*(3), 850–882.

Langley, A. (1999). Strategies for theorizing from process data. *Academy of Management Review*, *24*(4), 691–710.

Langley, A. (2007). Process thinking in strategic organization. *Strategic Organization*, *5*(3), 271–282.

Sandberg, J., Loacker, B., & Alvesson, M. (2015). Conceptions of process in organization and management. In H. Tsoukas, A. Langley, B. Simpson, & R. Garud (Eds.), *The emergence of novelty in organizations* (pp. 318–344). Oxford: Oxford University Press.

Thompson, M. (2011). Ontological shift or ontological drift? Reality claims, epistemological frameworks, and theory generation in organization studies. *Academy of Management Review*, *36*(4), 754–773.

Tsoukas, H., & Chia, R. (2002). On organizational becoming: Rethinking organizational change. *Organization Science*, *13*(5), 567–582.

Van de Ven, A. H., & Poole, M. S. (2005). Alternative approaches for studying organizational change. *Organization Studies*, *26*(9), 1377–1404.

DEMYSTIFYING CAQDAS: A SERIES OF DILEMMAS

Paula O'Kane

ABSTRACT

Computer-aided/assisted qualitative data analysis software (CAQDAS) supports qualitative and mixed methods researchers to organize, analyze, and explore data in a meaningful, and efficient, way. Successfully utilizing CAQDAS software can be challenging, particularly for the novice researcher. To assist all researchers 21 CAQDAS dilemmas are articulated. These relate to choosing, using, and getting started with the software, as well as writing about CAQDAS use. These dilemmas suggest there is no right way to use CAQDAS programs, rather the specific research project, along with researcher experience and philosophy, should drive the extent to which any project utilizes the extensive CAQDAS capabilities, while also encouraging the researcher(s) to drive their ideas and exploration beyond what they initially thought possible.

Keywords: CAQDAS; QDAS; Qualitative research; mixed methods; data analysis; coding

INTRODUCTION

As a qualitative researcher, my first introduction to computer-aided/assisted qualitative data analysis software (CAQDAS) was NUD*IST 6, a prequel to the current NVivo 12, and one of the most popular CAQDAS programs at the time. Its workings were rumored in my office of six PhD students from social science, but no one really knew what the program was or how it operated. Now, nearly 20 years on CAQDAS tools have become mainstream, but the mystic around these still lingers. This is often exemplified in the regularly CAQDAS workshops I organize, in which the reactions range from "*I wish I had known*

Advancing Methodological Thought and Practice
Research Methodology in Strategy and Management, Volume 12, 133–152

ISSN: 1479-8387/doi:10.1108/S1479-838720200000012020

about this software earlier" to "*this is all too over-whelming*," as we engage with querying and advanced features. The excitement of the basic tools, covered in the earlier session, often gives way to a feeling of uncertainty in later sessions.

The aim of this paper is not to provide definitive answers about the decision to use qualitative software, which package to use, or which projects are best suited to CAQDAS. Rather, it is to demystify the CAQDAS environment, to articulate the dilemmas we all face when using CAQDAS programs, and to reassure researchers that there is no one way to engage with the software. Each researcher and research team is unique, each project is different, and the operationalization of CAQDAS software will change from one project to the next. Therefore, the aim is to help you understand how the myriad of programs have been, and can be, used in practice for both the novel and the experienced user. In doing so, I explore the utility of current CAQDAS publications, although there are few within the strategy and management field. I discuss this around a series of contradictions, or dilemmas, which the qualitative researcher may face during their CAQDAS journey. These dilemmas help to frame both the potential of CAQDAS tools and the inherent difficulties of making these tools work in ways which maintain individual and qualitative integrity and allow researchers to explore and interpret their data in a meaningful and useful way (Gioia, 2019).

In this chapter, as I do at the end of each training session, I leave participants with three caveats. First, only use what is useful and relevant to your research. Just because the software can do it, does not mean you should do it. Second, know your research process, understand this and use this to guide your use of CAQDAS. Third, and to the other extreme, consider how CAQDAS tools and approaches can help you explore ideas and concepts in ways which would have been difficult without the power of the software.

My CAQDAS Journey

Everyone has their own CAQDAS journey, which influences his or her view and use of the software. As alluded to above, I am an NVivo user, and have been for nearly 20 years. I fell into my role as "expert" when the research graduate school in the University where I had completed my PhD asked if I could run NVivo training sessions. As a slightly naïve 25 year old, I agreed and have since run hundreds of sessions across institutions where I work, as an invited trainer, at conferences and workshops, and at one point as a QSR International (parent company of NVivo) trainer for Europe. It has been a journey of discovery, more often learning what I do not know than what I do know. While the experience of providing training has enabled me to become more familiar with the tools, I do not see myself as an expert. I am still learning and always will be.

Many scholars have written extensively about CAQDAS packages, a few of whom I have met over the years (although they probably do not remember meeting me). Anne Lewins and Christina Silver's comprehensive and in-depth book takes on the challenge of comparing and operationalizing the main CAQDAS software packages. Achieving this is impressive, and I highly recommend their book as a starting point (Silver & Lewins, 2014). Along with

Nicholas Woolf, Christina Silver has also developed the Five-Level QDA method, and they have adapted this for NVivo, Atlas.ti, and MAXQDA (Woolf & Silver, 2017a, 2017b, 2017c). Nigel Fielding, who instigated the CAQDAS networking project at Surrey University in the 1990s, and his coauthor Raymond Lee together coined the term CAQDAS in their early work (Fielding & Lee, 1991). Susanne Friese, who many years ago at a workshop I attended in Utrecht, asked the question – "how does CAQDAS influence/shape our qualitative research process?" This question has always intrigued me, and although I am still pondering the answer, I tend to support Bringer, Johnston, and Brackenridge's (2006, p. 263) assertion that "how well the researcher follows the chosen research methodology and applies the selected research methods will have a greater impact than whether CAQDAS is used." Susanne's book on using Atlas.ti in qualitative research is now in its third edition (Friese, 2019). Lyn Richards and her hilarious story of her child eating the paper cutouts of her "codes" not only influenced me but also led to the subsequent development of NUD*IST, together with her husband Tom, a computer programmer (Richards & Richards, 1998). More recently, Pat Bazeley has emerged as the NVivo expert. Pat and Kristi Jackson have just published the third edition of their NVivo guide (Jackson & Bazeley, 2019). It is also worthy of note that much of the early CAQDAS work came from the disciplines of nursing and social work. Both disciplines embraced qualitative research and software long before strategy and management, and they remain leaders in discussing software utility (see for example, Leitch, Oktay, & Meehan, 2016; Moylan, Derr, & Lindhorst, 2015).

CAQDAS Packages

There are seven main software packages specifically dedicated to qualitative data analysis and coding, as well as specific text analysis tools such as Leximamcer and Wordstat, but many qualitative researchers I meet still espouse the value of a simple excel spreadsheet or word document (Ose, 2016). I do not specifically recommend any package rather provide a resource to help the reader make their own decision about the CAQDAS program best suited to both them as an individual researcher and their current project. Four of these packages have been around since the late 1980s or early 1990s and continue to be market leaders (NVivo, Atlas.ti, QDA Miner, and MAXQDA). A newer entrant into the market with a solely cloud-based offering is Dedoose, popular with geographically dispersed teams. The remaining two packages, HyperRESEARCH and Transana, have less functionality and provide a low or no-cost option for researchers. Table 1 provides an overview of each of these tools, and a detailed review can be found in Silver and Lewins (2014). Before deciding on the package, the qualitative researcher in strategy and management first needs to consider if using CAQDAS is right for them.

The software dilemma: As happens often, my colleague recently received a review of her journal submission with the comment "why did you not use NVivo?" In my role as associate editor and NVivo trainer, I have encountered the same reviewer comment on numerous occasions. I have to admit, in my head I

Table 1. CAQDAS Programs.

CAQDAS Software	Latest Version	Platform	Introduction	History/Focus	Book
NVivo	v.12	Windows; Mac with reduced functionality	1990s	Grounded theory, inductive research.	Jackson and Bazeley (2019)
Atlas.ti	v.8 (plus cloud)	Windows; Mac	1980s	Hermeneutic text analysis; deep understanding of the data; grounded theory.	Friese (2019)
QDA Miner	V.5	Windows	1980s	Statistical and visualizations tools; part of suite of programs from Provalis.	n/a
MAXQDA	2018	Windows; Mac	1980s	Mixed methods research; visualizations.	Kuckartz and Rädiker (2019)
Dedoose	Cloud-based	n/a	2000s	Mixed methods research; supporting geographically disbursed teams; market disrupter.	Salmona, Lieber, and Kaczynski (2019)
Hyper-RESEARCH	v.4	Windows; Mac	1990s	Case-based analysis; integrated with transcription software; low sophistication; originally designed for Mac.	n/a
Transana	v.3	Windows; Mac	2000s	Low-cost, open source software, also used for transcribing.	n/a

blame ill-informed, often quantitative, reviewers who see CAQDAS as a panacea for solving the reliability, trustworthiness, or rigor problems they see with qualitative research. I wish I could thwart this attitude without seeing a rebuttal about the lack of rigor in qualitative research (Harley & Faems, 2017). Even as a firm believer in the use of software for qualitative research, these comments still frustrate me. While I do believe that CAQDAS, if well used, has the potential to increase trustworthiness, transparency, or rigor (Abramson, Joslyn, Rendle, Garrett, & Dohan, 2018; O'Kane, Smith, & Lerman, forthcoming), the software certainly should not be seen as the only way to achieve this.

Both seasoned and novice qualitative researchers struggle with the software dilemma, with many arguments for and against it. Those against it suggest it can be too cumbersome, distances the researcher, or invokes elements of the quantitative paradigm, which can sit uncomfortably; but these criticisms have lessened in recent years (Ose, 2016; Silver, 2018). Personally, my past research projects have been a mix of full, some, and no CAQDAS use. The reasons differ. Sometimes my colleagues who are leading the project do not want or know how to use CAQADS and so I follow their lead, whereas in other projects, colleagues want to learn the power of the software (such as our recent performance management research, Brown, O'Kane, Mazumdar, & McCracken, 2019). Other colleagues are firm believers in the software but often do not use it to its full capability. The decision can also be data driven. When we have a small in-depth dataset, it seems easier to work through this manually, sitting around a table, discussing as we go (as we did in our Work Futures Otago project, Walton, O'Kane, & Ruwhiu, 2019). I honestly have no agenda, I work with the needs of the project and my colleagues to either drive the use of CAQDAS or to take a step back and follow the needs of others. My one caveat is that I come to this decision from a place of experience. I know when I need to gently guide my colleagues into using CAQDAS software and when I do not. Just recently, in a new project, a colleague decided to engage with Dedoose. I carefully asked why Dedoose was chosen and was told it was because of the cloud-based capability for a geographically dispersed team. This sounded like a sensible reason, so I did not push for NVivo. Instead, I am relishing the challenge of learning Dedoose with this team. This leads to the second dilemma, which software package to use.

The CAQDAS software package dilemma: Perhaps, the hardest task for researchers considering CAQDAS is the decision surrounding which software package to adopt. The seven software packages I review have similar but different utility. This came to the fore when we were writing our recent article on transparency and trustworthiness in CAQDAS (O'Kane et al., forthcoming). While trying to compare just two software packages, specifically in the area of inductive qualitative analysis, we found the terminology used in each program could differ, and often even be contradictory. One of our aims in this paper was to create a common parlance that translated across software packages. This proved more difficult than we anticipated.

Although existing sources can be utilized in the decision about software adoption (see for example, Silver & Lewins, 2014), the reality is that most researchers predicate the decision on the package supported by their institution. If your research institution holds a site license for a CAQDAS package, this tends to be the determining factor as this reduces or eliminates the direct cost to the researcher or research project. Site licenses also create a community of practice around the CAQDAS software, which enable both informal and formal support. Unfortunately, site licenses often reinforce the value of the market leader, the organization with the best sales and marketing pitch! This is often country dependant; for example, the United Kingdom, the United States, Canada, and Australia tend to use NVivo primarily, whereas mainland European countries and some disciplines often use Atlas.ti (for example, Geography because of the GIS capability). Dedoose, as a subscription package, is disrupting this model as it provides a low-commitment alternative to the traditional site or individual license.

If your decision is not predicated on availability and cost, then the CAQDAS package used by other scholars or colleagues in your field or institution tends to be the next determining factor. Taking this approach provides a support mechanism when you inevitably need to ask some questions about how to complete an action in the software package. My door is often knocked upon by colleagues and PhD students, as well as emails from far and wide, asking how NVivo can be used to support their research, or simply trying to remember the purpose of a "case node." If I do not have the answer straightaway, I relish the challenge of working this out and continuing to learn elements of the package with which I am less familiar. Currently I am working through the link between Endnote and NVivo for literature reviews.

If neither of these factors are relevant to you or if your institution supports more than one CAQDAS program, then, as suggested above, a review of Silver and Lewins (2014) is a good starting point. Their book gives the reader a flavor of the different functionality of the CAQDAS programs and how these can be used in practice. Additionally, published articles may feature in-depth reflections on the ways in which researchers have used different software programs, and reading some of these can support understanding of the potential of CAQDAS generally and the different packages specifically (for example, Paulus & Lester's, 2016, use of Atlas.ti in a discourse analysis study; Hutchison, Johnston, & Breckon's, 2010, use of NVivo in a grounded theory project). Table 2 provides an overview of CAQDAS articles published since 2010 (sourced from an abstract and keyword search of Scopus, Business Source Complete, and Sage) and differentiates these by CAQDAS program and data analysis approach.

Only one of the 13 articles in Table 2 is from the field of strategy and management (Hajro, Gibson, & Pudelko, 2017). Therefore, I conducted a Scopus search to attempt to understand CAQDAS use in strategy and management research. An "All Fields" search (which includes article references, but not the full text) for CAQDAS terms (NVivo OR Altas.ti OR MAXQDA OR "QDA Miner" OR Dedoose OR CAQDAS OR QDAS) for the period 2010–2019 resulted in 2,645 journals results. By narrowing this to the field of Business, Management and Accounting, the number reduced to 1,054. In an attempt to understand more fully the spread across strategy and management journals, I further refined the results to those journals listed in the FT50 (see Table 3). These revealed only 20 articles, the majority of which appeared in the *Journal of Business Ethics*.

Although on the surface, this might raise concerns about whether strategy and management researchers are utilizing CAQDAS, it more likely results from one of two things. The Scopus search would not have located any articles with only brief mentions of CAQDAS or specific CAQDAS programs in the methodology (see, for example, Hajro et al., 2017, which did not appear in the search results). Additionally, it could be explained by a lack of space in journal articles to fully explain the methodological processes (see, for example, Mathias & Smith, 2016, in *Organizational Research Methods* and Jarzabkowski, Bednarek, & Cabantous, 2015, in *Human Relations*). This is something I too am guilty of in my research and was one of the driving forces behind our recent article, i.e., making it easier to communicate CAQDAS use in qualitative papers (O'Kane et al., forthcoming).

The majority of articles in Table 2 takes an interpretive approach, whether this be grounded theory, discourse analysis, or ethnography, among others. This

Table 2. Articles Reviewing CAQDAS Use.

Author	Title	Journal	Article Type	Data Management	Coding	Exploration	Visualization	Reflexivity	Verification
O'Kane et al. (forthcoming)	Building transparency and trustworthiness in inductive research through Computer-Aided Qualitative Data Analysis Software	*Organizational Research Methods*	In-depth review of NVivo and QDA Miner to support transparency and trustworthiness		●	●	●	●	●
Geisler (2018)	Coding for language complexity: The interplay among methodological commitments, tools, and workflow in writing research	*Written Communication*	Compares Excel, MAXQDA, and Dedoose for coding language complexity	●	●	●		●	
Abramson et al. (2018)	The promises of computational ethnography: Improving transparency, replicability, and validity for realist approaches to ethnographic analysis	*Ethnography*	Uses Atlas.ti for an ethnographic study				●		●
Hajro et al. (2017).	Knowledge exchange processes in multicultural teams: Linking organizational diversity climates to teams' effectiveness	*Academy of Management Journal*	Uses Atlas.ti to support case study analysis of interviews, observations, and public documentation	●	●	●		●	●
Paulus and Lester (2016)	ATLAS.ti for conversation and discourse analysis studies	*International Journal of Social Research Methodology*	In-depth review of an Atlas.ti project using discourse and conversation analysis	●	●	●		●	
Carcary (2011)	Evidence analysis using CAQDAS: Insights from a qualitative researcher	*The Electronic Journal of Business Research Methods*	In-depth review of an NVivo project using interpretative case studies	●	●	●	●	●	
Hoover and Koerber (2011)	Using NVivo to answer the challenges of qualitative research in professional	*IEEE Transactions on*	A tutorial on how and why to use NVivo	●					

Table 2. (*Continued*)

Author	Title	Journal	Article Type	Data Management	Coding	Exploration	Visualization	Reflexivity	Verification
	communication: Benefits and best practices	*Professional Communication*							
Housley and Smith (2011)	Telling the CAQDAS code: Membership categorization and the accomplishment of 'coding rule' in research team talk	*Discourse Studies*	In-depth review of an Atlas.ti project using a deductive ethnomethodological approach		●			●	●
Mavrikis and Geraniou (2011)	Using qualitative data analysis software to analyse students' computer-mediated interactions: The case of MiGen and Transana	*International Journal of Social Research Methodology*	In-depth review of a Transana project coding video data using a grounded theory approach		●				
Sinkovics and Penz (2011)	Multilingual elite-interviews and software-based analysis: Problems and solutions based on CAQDAS	*International Journal of Market Research*	In-depth review of an NVivo project using an interpretive, grounded theory approach	●	●	●	●		
Hutchison et al. (2010)	Using QSR-NVivo to facilitate the development of a grounded theory project: an account of a worked example	*International Journal of Social Research Methodology*	In-depth review of an NVivo project using grounded theory	●	●	●	●	●	●
King (2010)	'Membership matters': Applying Membership Categorisation Analysis (MCA) to qualitative data using Computer-Assisted Qualitative Data Analysis (CAQDAS) Software	*International Journal of Social Research Methodology*	In-depth review of an NVivo project using membership categorization analysis	●	●	●		●	
Bringer et al. (2006)	Using Computer-Assisted Qualitative Data Analysis [illegible]	*Field Methods*	In-depth review of an NVivo project using [illegible]	●	●	●	●	●	●

Table 3. CAQDAS Coverage in FT50 *Strategy and Management Journals* (2010: 2019 Inclusive).

FT 50 Journal	n = 20
Academy of Management Journal	0
Academy of Management Review	0
Administrative Science Quarterly	0
Entrepreneurship Theory and Practice	1
Harvard Business Review	0
Human Relations	2
Human Resource Management	2
Journal of Business Ethics	8
Journal of Business Venturing	0
Journal of International Business Studies	2
Journal of Management	0
Journal of Management Studies	1
Management Science	0
Organization Science	0
Organization Studies	0
Organizational Behavior and Human Decision Processes	0
Research Policy	2
Sloan Management Review	0
Strategic Entrepreneurship Journal	2
Strategic Management Journal	0

might suggest that interpretative analysis lends itself better to the use of CAQDAS. An alternative explanation is that interpretative projects provide good fodder for writing an in-depth review of CAQDAS use, when compared to deductive projects. As can be seen in Table 2, most reviews discuss the use of one CAQDAS package, while a minority compare programs to each other. Interestingly, NVivo features in 7 of the 13 articles, reflecting its position as market leader, but this should not be taken as an endorsement that it is the superior package. To better understand these trends and how researchers are actually using CAQDAS tools, it would be useful to conduct a content analysis of the method sections of strategy and management papers.

Early CAQDAS packages were designed with different qualitative research approaches in mind. For example, NVivo was originally designed to support "grounded theory techniques of open coding, writing memos, axial coding and creating models" (Bringer et al., 2006, p. 245), and the majority of the articles in Table 2 take this approach. An in-depth review of long-term physical activity change, also using grounded theory, concluded, "QSR-NVivo is a powerful tool that, if used appropriately, can facilitate many aspects of grounded theory" (Hutchison et al., 2010, p. 283). CAQDAS and grounded theory continue to be viewed as excellent partners, particularly Atlas.ti and NVivo. Carcary (2011) discusses her use of NVivo for concept creation and management, cross-case tabulation, and modeling to reach new theoretical insights and displays actual examples from the NVivo project in the evaluation of a new student ICT administrative system.

Similarly, Atlas.ti tends to be more inductive, but as can be seen in Table 2, fewer articles have reflected in-depth upon the use of Atlas.ti. One exception is Paulus and Lester's (2016) review of utilizing CAQDAS in two different conversation and discourse analysis studies. One consisted of therapy data (therapy sessions and interviews with parents and clinicians from a pediatric clinic), with the aim to understand how autism was constructed. The other was an analysis of blog posts from an instructional task for nutrition students.

On the other hand, QDA Miner started as a more deductive tool and has a sister program Wordstat, designed for text analysis. To the best of my knowledge, the only in-depth review article utilizing QDA Miner is O'Kane et al. (forthcoming). MAXQDA was designed to support visualization of data, but many of the other programs now also provide extensive visualization tools. Giesler (2018) compares MAXQDA and Dedoose with tools like Excel. Both these CAQDAS programs were designed to support mixed methods research.

Mixed methods researchers appears to have been attracted to CAQDAS from early on with Fielding and Cisneros-Puebla (2009) exploring the use geographic information systems (GIS) data and combining this with qualitative social science CAQDAS tools, demonstrating the integration of different disciplines within the CAQDAS sphere. A similar review by Jung and Elwood (2010) talks about the specific discipline of CAQ-GIS (computer-aided qualitative geographic information systems). Fielding (2012) also points to the increased ability for CAQDAS tools to support quantitative and qualitative data integration and support data triangulation. Other methodological approaches include using CAQDAS for membership categorization analysis (King, 2010) and ethnography (Abramson et al., 2018).

The Mac dilemma: Tied to the software dilemma is the functionality of CAQDAS tools within the Mac environment. Many of the programs operate with reduced functionality on Mac (see Table 1). When I see trainees arrive in my office or at my training sessions with MacBooks, I invariably release a quiet sigh. NVivo, in particular, does not provide the same functionality to Mac users, and it is only in more recent times that the Windows and Mac version of NVivo have been able to "speak" to each other. Often Mac users have to convert their files to Windows versions of NVivo to delve further into the data, which can prove frustrating, particularly in team-based work. Cloud-based integration, such as that already seen in Dedoose and Atlas.ti, has the potential to remove this barrier for Mac users.

The software convergence dilemma: One of the biggest caveats I have about CAQDAS software is the continual development of new features (Humble, 2019). Although I see potential in the conversations and ideas these features create, they are particularly overwhelming for the novice researcher. Each program appears to be continually introducing increased, and often similar, functionality when the average researcher often only needs to code and review; what Richards (2004) described as "coding for retrieval" as opposed to "coding for query." This was identified by QSR International when they introduced three different options for NVivo 11 (starter, pro, and plus) following feedback from clients (me included). Given that this changed again with the introduction of NVivo 12 (starter was removed), I am not sure introducing different versions was particularly successful.

Conversely, Silver and Woolf (2019, p. 3) bemoan the "superficial" use of CAQDAS, and suggest that it "often has the effect of supressing some of the emergent aspects of analysis." I disagree. There is space to use CAQDAS in the way that is right for you as a user, but I also agree that there can be ignorance around the power of these programs and how they can support your research projects in useful ways which would be difficult or impossible without the software. More recently, the Rotterdam Exchange Format Initiative has brought the main software developers together to create an open source exchange format, to enable projects to operate in different software packages (Humble, 2019).

The team project dilemma: It is increasingly common for research projects to be team-based (many of the reviews in Table 2 discuss team projects). Having a number of researchers coding data can support interpretation and trustworthiness (Boyatzis, 1998). When using CAQDAS programs, it is essential to establish teamwork protocols to ensure the data are being treated and coded similarly by different researchers. First, establishing team member roles ensures everyone knows what they can and cannot do with the data and coding process (Jarzabkowski et al., 2015). Generally, it is good practice to have a lead researcher and the remainder of the team are coders. The lead researcher is responsible for maintaining the master copy, merging coding, keeping a record of how the coding and analysis process evolves, and selecting project parameters. Establishing clear project parameters are essential to a clear coding process (Hajro et al., 2017; Hoover & Koerber, 2011; Paulus & Lester, 2016). These include the level of freedom each coder will have to change aspects of the project, including other team members' codes, adding new codes or removing codes, the unit of coding the team will adopt (letters, words, sentences, or paragraphs), and whether each chunk of code can be coded more than once. Some CAQDAS facilitate the enforcement of project parameters, while others will require off-line agreement. More recently, NVivo introduced a specific package for working in teams. Memoing tools and the ability to merge multiple versions of the CAQDAS file can support teams to efficiently manage larger datasets (Paulus & Lester, 2016).

Once you have made the decision about which CAQDAS package to use and how your team will work together, you can then begin to delve into the software use.

USING CAQDAS IN PRACTICE

Using the articles reviewed in Table 2, this section explores how some research projects, as articulated in published papers, have utilized CAQDAS. These papers tended to discuss the use and utility of CAQDAS in relation to six key areas: data management, coding, exploration, visualization, reflexivity, and verification (see Table 2 for a review of which papers discuss each area).

Data Management

Lyn and Tom Richards first designed NVivo for data management purposes, and I still see this as the key tenant of CAQDAS tools. Managing data in a systematic way, keeping it all together in one place, and being able to search and retrieve information across multiple pieces of data, in different ways in seconds, is a very powerful function of all CAQDAS tools; nearly every article in Table 2 alludes to this "code and retrieve" functionality. Hoover and Koerber (2011) split data management into three areas – multiplicity, efficiency, and transparency – each of which are discussed below.

The dilemma of multiplicity: Keeping all your data in one place, that is within the CAQDAS environment, enables the research team to efficiently and quickly access and search multiple data types such as photos (with descriptions), video and audio (with transcripts), PDFs (such as adding literature in a grounded theory approach), and other data. Subsequently all coded data, regardless of type, can be viewed together within the code, making reviewing the code content easier and ensuring that a fuller picture of the data is available. This can ultimately support the researcher to offer richer insights.

The dilemma of efficiency: Most CAQDAS packages support multiple levels of organization (through, for example, folders, assigning attributes, or creating cases) which can chunk the data, enable it to be analyzed in different ways, and answer different research questions. This is perhaps one of the greatest strengths of the packages – making overwhelming amounts of data feel manageable. This, in turn, allows researchers to ensure all their project data have a voice. However, at the same time, it can be easy to "misorganize" the data. Understanding the CAQDAS program, creating protocols and working in teams, can help to alleviate this.

The dilemma of transparency: CAQDAS has been espoused as a way for qualitative researchers to increase transparency (O'Kane et al., forthcoming; Wickham & Woods, 2005). CAQDAS can support transparency through decision and action tracking, thus enabling researchers to accurately, and quickly, write the research methods sections of their papers. Decisions and actions can be captured using three methods: the CAQDAS log; researcher notes; and project file and node structure backups. The CAQDAS log automatically tracks all activities in the project and identifies detailed information such as the team member who made the change, date of change, and the specific commands, and changes made. I would caution relying on these as your main source of an audit or decision trail. Instead, I suggest using these to support your own accounts of the turns your qualitative analysis journey took.

Researcher notes (including memos) provide quick access to catch a fleeting thought, copy a portion of text for further comment, or capture other insights during the coding process, similar to a word file or notebook kept by many inductive researchers. This discipline of writing a journal of activities is one of the most important actions for qualitative researchers and CAQDAS users. Every day (or hour, or after a substantive activity) write down what you did and what you were thinking. This allows you to revisit the rationale for a search or for a reordered coding structure. Annotations to specific data chunks can be made in some programs, allowing ideas to be recorded beside the data.

Capturing project versions, through backups at key transitions points in the coding evolution, enables the researcher to return to a point in time if their thinking has taken them down a rabbit warren. Ideas can be explored without the worry of previous ideas being lost. Creating copies of code sequences, project files, or searches undertaken within the programs can also aid the sense-making process of disentangling thought processes.

Coding

The process of coding is inherently difficult to define and not easily replicated. It can be data-driven, theory-driven, or structural (Boyatzis, 1998), but whichever approach is taken, all CAQDAS programs are designed to support this process. Every software package has coding at its core, but critics of CAQDAS for interpretative analysis suggest that the software can stifle thinking though creating distance from the data and forcing hierarchical structures (Jackson, Paulus, & Woolf, 2018), while proponents often discuss the ability to iterate coding and demonstrate coding development (for example, Sinkovics & Penz, 2011).

The dilemma of distance: For some, getting too close to the data is an issue which can result in overcoding and make it particularly difficult to see patterns. While others suggest that CAQDAS programs create distance from the data as you cannot touch and feel it, the reality is that both can be countered by making good choices as a researcher. At times, I step away from the program and map ideas on paper, while at other times creating more codes allows me to see the data in another way. I particularly like the ability to iterate coding, and keep a track of this, whether you are working as an individual or in a team of researchers. This allows ideas to evolve. Visualization and modeling (discussed below) can support the researcher to oscillate between to distance and closeness within a CAQDAS project.

The dilemma of hierarchical structures: CAQDAS tools can nudge the researcher into viewing their codes in a hierarchical way, or following a boilerplate template, from the beginning of the coding process (Lê, Smith, Crook, & Boyd, 2019). This can be seen to stifle thinking, but as with distance, this is easily overcome by the researcher maintaining their integrity and using the tools in a way which suits their research paradigm. Rather than forcing structure, CAQDAS can enable, and support, reflection and exploration (see below), allowing our thinking to go from A to B via C, D, E, and F, supporting the qualitative research journey. For example, in NVivo, "free nodes" (with no hierarchy) can be utilized in the first round of open coding, and relationships can be used to link codes in differing ways.

Exploration

CAQDAS allows the researcher to explore data in different ways. For example, simply cross-tabulations on large datasets enable the researcher to understand if there are demographic or organizational differences. Text analysis and searching or querying abilities can bring the data to life. The dilemma here, in my experience, is that querying can get difficult to understand and can lead researchers down "rabbit holes" which can be difficult to get out of. I have also seen searches carried out inappropriately and results interpreted wrongly, giving false findings. These are not deliberate on the researchers' part but rather a function of not fully understanding the program.

The dilemma of overexploring: CAQDAS can support researchers to see the relationships among codes or to substantiate emerging connections. O'Kane et al. (forthcoming) provide a detailed review of different ways in which data can be explored using CAQDAS tools. The caution here is that these capabilities can also lead the researcher down different ways of thinking and away from the original scope of the research. On the one hand, I caution against the overuse of CAQDAS tools and features, but, on the other hand, I embrace the exploration of data. The energy expanded in collecting qualitative data is often huge: locating and accessing participants, transcribing data, reading and rereading text, and exploring unfruitful avenues. Being able to use these data in different ways, and enabling ideas to emerge, can, if done well, create new insights.

The dilemma of quantifying qualitative data: The qualitative research community are often divergent in their opinion around quantifying qualitative data, but there is no doubt CAQDAS has made this easier and more accurate (Jackson & Bazeley, 2019). As with most aspects of coding and exploration, reminding yourself of the research objectives and ensuring that you use appropriate quantitative measures to explain these can, in the right circumstances, provide additional insights into the data. With mixed methods research, complementing survey data with qualitative responses can be very powerful, and many CAQDAS programs support this (Fielding, 2012). Personally, I quantify in some projects and not in others.

The dilemma of workarounds: There is a contradiction in selecting the "right" CAQDAS tool. Although, on the one hand, software is converging (see above), on the other, each program has its unique functionality, and at times exporting to programs such as Excel or Word is useful. I term such techniques "workarounds." With the advent of open source exchange format between CAQDAS programs, if a specific function is not available then a workaround might be available in another program (Geisler, 2018). This adds a layer of opportunity and complexity to our analysis and exploration processes.

Visualization

Diagrams can be used at different points in the research process "as artefacts to stimulate discussion in interviews, assisting the researcher in formulating ideas, refining conceptualizations in the process of theory building and communicating ideas to others" (Buckley & Waring, 2013, p. 148).

The dilemma of diagrams: CAQDAS programs enable researchers to represent their coding structures, initial ideas, or potential relationships in diagrammatic format, which can be both dynamic (updating and changing as coding continues) and static (Hutchison et al., 2010). MAXQDA, in particular, was designed with this in mind. Sinkovics and Penz (2011) used models at both the conceptualization phase of their team project and also in the analysis to highlight the continuous iterations. Similarly, Bringer et al. (2006) used the modeling feature in NVivo to create distance from the data and enable them to create clusters by moving the visual representations of the node around. I am a fan of models and diagrams to represent ideas and concepts in qualitative research, but I prefer to use pen and paper, as I found the software awkward and clunky. Many other researchers find this one of the most useful aspects of CAQDAS tools (see for example, Abramson et al., 2018). The power now available in today's CAQDAS programs have taken visual exploration to another level, utilizing many text analysis concepts to produce heat maps, word clouds, or cluster analysis (see, for instance, Hutchison et al., 2010).

Reflexivity

CAQDAS really comes to life when we consider the opportunity the programs give us to reflect upon our data, to view it in different ways, and to take our thinking to another level. Many of the ideas above support reflexivity: subsets of data produced from queries or questions of our data (exploration); word clouds visually demonstrating frequently occurring words or more sophisticated tools such cluster analysis (visualization); or notes (memos) made while creating and connecting codes to the text (transparency). Memos, in particular, are often synonymous with reflexivity particularly from a grounded theory perspective. Hutchison et al. (2010) provide an excellent overview of five different ways in which memos might be utilized in the CAQDAS environment (research diary, reflective, conceptual, emergent questions, and explanatory).

The dilemma of reflecting too much: Reflection can happen in two different ways. It can be driven from our research question, flowing naturally from our review of the literature or it can happen organically, in the process of doing research. We might be having "an explore" of the data, meandering through it and using our subject knowledge and expertise to reflect. When we use CAQDAS, another layer is added. It provides the opportunity to explore those ideas with different tools (searching, querying, and so on) and to see new ones – to understand our data in another way – to take our thinking to a higher level. This does not negate the need for good and strong research questions, but it does allow us to understand different connections and concepts, to evolve new ideas, new theory and different ways of viewing the data we have often painstakingly worked to gather. Most of the articles reviewed in Table 2 use reflexivity either explicitly or implicitly through other types of exploration and visualization.

Verification

As many researchers call for increased trustworthiness in qualitative research, the verification functionality CAQDAS can offer becomes invaluable. In this paper, I do not get into a debate about the necessity of verification, instead I reflect on recent papers which help to explain how CAQDAS can support researchers who wish to verify their analysis.

The trustworthiness dilemma: O'Kane et al. (forthcoming) provide an in-depth review of 11 different ways in which CAQDAS tools can be utilized to increase trustworthiness. Some of this is supported by the integration of text analytics into some CAQDAS packages. While this is a positive step and can be a great source for exploring coding consistency, it can also be problematic as CAQDAS text analysis functions are not as effective as the specific tools designed for this. Trustworthiness can also be established by segmenting and comparing data across cases or groups within a project (see for example, Hajro et al., 2017) or using numerical analysis to establish coding consistency within team projects or to support coding trustworthiness (Housley & Smith, 2011).

The interrater reliability dilemma: Many projects use multiple coders for a variety of reasons from splitting large datasets to double coding each data segment. Researchers can use a range of approaches to support consistency between coders, or over time for an individual coder: comparison and discussion; calculating agreement and disagreement between coders; or computing interrater reliability (IRR, using Cohen's kappa coefficient, Fleiss's kappa statistic, Pearson product–moment correlation coefficient, or Spearman's rank correlation coefficient, Miles & Huberman, 1994; Saldaña, 2009). The IRR debate in qualitative research is a controversial one. Many researchers disagree completely with using these statistics, others use them simply to support their own coding, while at the other end others report them in their findings (Campbell, Quincy, Osserman, & Pedersen, 2013). Personally, the jury is out for me. I tend not to report IRR, but this may change.

GETTING STARTED WITH CAQDAS

Data collection or software first dilemma: The software decision is made and now what? How do you go about successfully working with the software? Unfortunately, there is no right answer to this. After years of running introductory CAQDAS training sessions, nearly everyone has one of these two reactions. The first, when undertaking training early in the qualitative research journey, often with no data and limited understanding of qualitative analysis, the rationale for using the software can be difficult to see. The trainee is learning with an unfamiliar dataset making the process even more abstracted. I have seen these trainees complete the training again later in their research journey. This reaction therefore tends to be a feeling of being overwhelmed and unsure about how to make best use of the software.

The other reaction, from those who have begun or completed their qualitative projects is "*I wish I had known this sooner.*" They see different ways they could have organized their data and shortcuts they could have made use of to ease the coding and analysis process. Probably the most successful groups I have seen are teams of researchers training together for a project they are about to begin. This reinforces the value of communities of practice and support from colleagues. Together, they discuss and question how they might use the software. They get less lost in the potential and focus more on the reality of their projects.

So to answer the question, there really is not an ideal time to learn the software, and it often just becomes a matter of when the training is available. Running one-to-one sessions has proven better for some people, but for most, the cost is out of reach. What I do know is that even for those with experience of CAQDAS, there is always something new they learn in a course, and conversely, I often learn something new from my participants when they challenge me to think differently about the software.

The dilemma of getting started: Back to the question at hand, how do you get started? Practice and play, create backups of the data, talk to other people who use the software. I am a great believer in a community of practice. Thinking back, that one computer in our PhD office with NUD*IST loaded, might have been the best thing that could have happened for my learning. It opened up the avenues of discussion about how each person was using the software, what they were learning and the different ways they were using it. Dialogue still happens, monthly I get a phone call asking me how to do something from a colleague or a past participant in my training programs – often I can answer this quite quickly, other times it sets me a challenge (sometimes a distraction) to work out how to do what they are proposing. I learn too.

The dilemma of data preparation: The preparation of the data, particularly text, for importing is the first step in an effective coding procedure. I advise finalizing the data before importing. Although many software packages now allow for transcription within their environments, I still find it cleaner to use the more powerful word processing tools to ensure the data are accurate. This includes a thorough spelling and accuracy check.

Newer ways of preparing data are emerging. For example, later software versions (such as NVivo and Atlas.ti) are enabling what Parameswaran, Ozawa-Kirk, and Latendresse (forthcoming) describe as "live coding," coding directly from video or audio. In their comparison between live and transcript coding, they suggest that live coding allowed them "to see and hear the participants… which allowed intent, context, and meaning of the words to be present in the results." Similarly, Mavrikis and Geraniou (2011) used Transana for video coding and reflected on the importance of context which they established through coding nonverbal behavior. A compromise can be to transcribe the data with timestamps and subsequently listen to the recording or watch the video as you read the transcription (Paulus & Lester, 2016). This could be seen as the best of both worlds. Chandler, Anstey, and Ross (2015) go even further when they advocate for hypermodal dissemination (embedding audio clips in presentations or in written manuscripts). However, the ethics of disclosing participant voice in this scenario are inherently problematic and may outweigh the potential benefits.

CONCLUDING THOUGHTS: MOVING THE FIELD FORWARD

Although the series of dilemmas presented above may reflect, on the surface, a field of study in chaos, this is not true. We are developing and growing, we have a wonderful opportunity to highlight good practice in the use of CAQDAS in our research and to provide readers with useful information about the CAQDAS journey we have used in our own projects. This requires time and effort, an understanding of how CAQDAS has influenced our research, a way to articulate how we have used the tools available, and strong guidance and practice for writing the research methods sections of our papers to incorporate the CAQDAS journey.

The dilemma of space: This is the grand challenge of qualitative research, how do we describe what we did in enough detail for the reviewer to understand that we were comprehensive in our analysis, yet stay within the word count dictated by the journal. We then begin to shortcut what we actually did, processes which took months, and required iteration upon iteration of analysis. I am guilty of shortcutting, often compressing my research process into three stages of analysis, choosing the three most distinctive points my research process. In this vein, I call for more time and space to be dedicated to the role of software in the research process, which I contend can be achieved by utilizing the potential of online appendices to fully articulate our qualitative analysis processes.

REFERENCES

Abramson, C. M., Joslyn, J., Rendle, K. A., Garrett, S. B., & Dohan, D. (2018). The promises of computational ethnography: Improving transparency, replicability, and validity for realist approaches to ethnographic analysis. *Ethnography, 19*(2), 254–284.

Boyatzis, R. E. (1998). *Transforming qualitative information: Thematic analysis and code development*. Thousand Oaks, CA: SAGE Publications.

Bringer, J. D., Johnston, L. H., & Brackenridge, C. H. (2006). Using Computer-Assisted Qualitative Data Analysis Software to develop a grounded theory project. *Field Methods, 18*(3), 245–266.

Brown, T. C., O'Kane, P., Mazumdar, B., & McCracken, M. (2019). Performance management: A scoping review of the literature and an agenda for future research. *Human Resource Development Review, 18*(1), 47–82.

Buckley, C. A., & Waring, M. J. (2013). Using diagrams to support the research process: Examples from grounded theory. *Qualitative Research, 13*(2), 148–172.

Campbell, J. L., Quincy, C., Osserman, J., & Pedersen, O. K. (2013). Coding in-depth semistructured interviews: Problems of unitization and intercoder reliability and agreement. *Sociological Methods & Research, 42*(3), 294–320.

Carcary, M. (2011). Evidence analysis using CAQDAS: Insights from a qualitative researcher. *Electronic Journal of Business Research Methods, 9*(1), 10–24.

Chandler, R., Anstey, E., & Ross, H. (2015). Listening to voices and visualizing data in qualitative research: Hypermodal dissemination possibilities. *SAGE Open, 5*(2). Retrieved from https://journals.sagepub.com/doi/full/10.1177/2158244015592166

Fielding, N., & Cisneros-Puebla, C. A. (2009). CAQDAS-GIS convergence: Toward a new integrated mixed method research practice? *Journal of Mixed Methods Research, 3*(4), 349–370.

Fielding, N., & Lee, R. (1991). *Using computers in qualitative research*. London: SAGE Publications.

Fielding, N. G. (2012). Triangulation and mixed methods designs: Data integration with new research technologies. *Journal of Mixed Methods Research, 6*(2), 124–136.

Friese, S. (2019). *Qualitative data analysis with ATLAS.ti*. Los Angeles, CA: SAGE Publications.

Geisler, C. (2018). Coding for language complexity: The interplay among methodological commitments, tools, and workflow in writing research. *Written Communication*, *35*(2), 215–249.

Gioia, D. A. (2019). If I had a magic wand: Reflections on developing a systematic approach to qualitative research. In B. Boyd, T. R. Crook, J. K. Lê, & A. D. Smith (Eds.), *Standing on the shoulders of giants* (Vol. 11, pp. 27–37). Research Methods in Strategy and Management. Bingley: Emerald Publishing.

Hajro, A., Gibson, C. B., & Pudelko, M. (2017). Knowledge exchange processes in multicultural teams: Linking organizational diversity climates to teams' effectiveness. *Academy of Management Journal*, *60*(1), 345–372.

Harley, B., & Faems, D. (2017). Theoretical progress in management studies and the role of qualitative research. *Journal of Management Studies*, *54*(3), 366–367.

Hoover, R. S., & Koerber, A. L. (2011). Using NVivo to answer the challenges of qualitative research in professional communication: Benefits and best practices tutorial. *IEEE Transactions on Professional Communication*, *54*(1), 68–82.

Housley, W., & Smith, R. J. (2011). Telling the CAQDAS code: Membership categorization and the accomplishment of 'coding rules' in research team talk. *Discourse Studies*, *13*(4), 417–434.

Humble, A. (2019). Computer-Aided Qualitative Analysis Software. Retrieved from https://methods.sagepub.com/foundations/computer-aided-qualitative-analysis-software

Hutchison, A., Johnston, L., & Breckon, J. (2010). Using QSR-NVivo to facilitate the development of a grounded theory project: An account of a worked example. *International Journal of Social Research Methodology*, *13*(4), 283–302.

Jackson, K., & Bazeley, P. (2019). *Qualitative data analysis with NVivo*, Los Angeles, CA: SAGE Publications.

Jackson, K., Paulus, T., & Woolf, N. H. (2018). The walking dead genealogy: Unsubstantiated criticisms of Qualitative Data Analysis Software (QDAS) and the failure to put them to rest. *Qualitative Report*, *23*(13), 74–91.

Jarzabkowski, P., Bednarek, R., & Cabantous, L. (2015). Conducting global team-based ethnography: Methodological challenges and practical methods. *Human Relations*, *68*(1), 3–33.

Jung, J.-K., & Elwood, S. (2010). Extending the qualitative capabilities of GIS: Computer-Aided Qualitative GIS. *Transactions in GIS*, *14*(1), 63–87.

King, A. (2010). 'Membership matters': Applying Membership Categorisation Analysis (MCA) to qualitative data using Computer-Assisted Qualitative Data Analysis (CAQDAS) Software. *International Journal of Social Research Methodology*, *13*(1), 1–16.

Kuckartz, U., & Rädiker, S. (2019). *Analyzing qualitative data with MAXQDA: Text, audio, and video*. Cham: Springer International Publishing.

Lê, J. K., Smith, A. D., Crook, R., & Boyd, B. (2019). Why research methodology in strategy and management remains as important as ever. In B. Boyd, R. Crook, J. K. Lê, & A. D. Smith (Eds.) & K. B. Brian (Trans.), *Standing on the shoulders of giants: Traditions and innovations in research methodology* (Vol. 11, pp. 1–13). Bingley: Emerald Publishing.

Leitch, J., Oktay, J., & Meehan, B. (2016). A dual instructional model for Computer-Assisted Qualitative Data Analysis Software integrating faculty member and specialized instructor: Implementation, reflections, and recommendations. *Qualitative Social Work*, *15*(3), 392–406.

Mathias, B. D., & Smith, A. D. (2016). Autobiographies in organizational research: Using leaders' life stories in a triangulated research design. *Organizational Research Methods*, *19*(2), 204–230.

Mavrikis, M., & Geraniou, E. (2011). Using Qualitative Data Analysis Software to analyse students' computer-mediated interactions: The case of MiGen and Transana. *International Journal of Social Research Methodology*, *14*(3), 245–252.

Miles, M. B., & Huberman, A. M. (1994). *Qualitative data analysis: An expanded sourcebook*. Thousand Oaks, CA: SAGE Publications.

Moylan, C. A., Derr, A. S., & Lindhorst, T. (2015). Increasingly mobile: How new technologies can enhance qualitative research. *Qualitative Social Work*, *14*(1), 36–47.

O'Kane, P., Smith, A., & Lerman, M. P. (forthcoming). Building transparency and trustworthiness in inductive research through Computer-Aided Qualitative Data Analysis Software. *Organizational Research Methods*.

Ose, S. O. (2016). Using Excel and Word to structure qualitative data. *Journal of Applied Social Science*, *10*(2), 147–162.

Parameswaran, U. D., Ozawa-Kirk, J. L., & Latendresse, G. (forthcoming). To live (code) or to not: A new method for coding in qualitative research. *Qualitative Social Work*.

Paulus, T. M., & Lester, J. N. (2016). ATLAS.ti for conversation and discourse analysis studies. *International Journal of Social Research Methodology*, *19*(4), 405–428.

Richards, T. (2004). Retrieve vs. query – What is the purpose of coding. In *CAQDAS conference proceedings*. Brisbane: Griffith University.

Richards, T. J., & Richards, L. (1998). Using computers in qualitative research. In N. K. Denzin & Y. S. Lincoln (Eds.), *Collecting and interpreting qualitative materials* (pp. 445–462). London: SAGE Publications.

Saldaña, J. (2009). *The coding manual for qualitative researchers*. Thousand Oaks, CA: SAGE Publications.

Salmona, M., Lieber, E., & Kaczynski, D. (2019). *Qualitative and mixed methods data analysis using Dedoose: A practical approach for research across the social sciences*. Thousand Oaks, CA: SAGE Publications.

Silver, C. (2018). CAQDAS at a crossroads: Choices, controversies and challenges. In A. P. Costa, L. P. Reis, F. N. d. Souza, & A. Moreira (Eds.), *Computer supported qualitative research: Second international symposium on qualitative research (ISQR 2017)* (pp. 1–13). Cham: Springer International Publishing.

Silver, C., & Lewins, A. (2014). *Using software in qualitative research: A step-by-step guide*. Thousand Oaks, CA: SAGE Publications.

Silver, S., & Woolf. (2019). Five-level QDA method. SAGE Research Methods Foundations. Retrieved from https://methods.sagepub.com/foundations/five-level-qda-method

Sinkovics, R. R., & Penz, E. (2011). Multilingual elite-interviews and software-based analysis: Problems and solutions based on CAQDAS. *International Journal of Market Research*, *53*(5), 705–724.

Walton, S., O'Kane, P., & Ruwhiu, D. (2019). Developing a theory of plausibility in scenario building: Designing plausible scenarios. *Futures*, *111*, 42–56.

Wickham, M., & Woods, M. (2005). Reflecting on the strategic use of CAQDAS to manage and report on the qualitative research process. *Qualitative Report*, *10*(4), 687–702.

Woolf, N. H., & Silver, C. (2017a). *Qualitative analysis using NVivo: The five-level QDA® method*. Abingdon: Taylor & Francis.

Woolf, N. H., & Silver, C. (2017b). *Qualitative analysis using ATLAS.ti: The five-level QDA® method*. Abingdon: Taylor & Francis.

Woolf, N. H., & Silver, C. (2017c). *Qualitative analysis using MAXQDA: The five-level QDATM method*. Abingdon: Taylor & Francis.

INDEX

Printed in the United States
By Bookmasters